PHYSICAL CHEMISTRY OF FOOD PROCESSES

VOLUME I

Fundamental Aspects

PHYSICAL CHEMISTRY OF FOOD PROCESSES
VOLUME I

Fundamental Aspects

Edited by

Ion C. Baianu

Associate Professor
Physical Chemistry of Foods
University of Illinois at Urbana

An **avi** Book
Published by Van Nostrand Reinhold
New York

An AVI Book
(AVI is an imprint of Van Nostrand Reinhold)

Copyright © 1992 by Van Nostrand Reinhold
Library of Congress Catalog Card Number 91-46028
ISBN 0-442-00580-6

Printed in the United States of America.

Van Nostrand Reinhold
115 Fifth Avenue
New York, New York 10003

Chapman and Hall
2-6 Boundary Row
London, SE1 8HN, England

Thomas Nelson Australia
102 Dodds Street
South Melbourne 3205
Victoria, Australia

Nelson Canada
1120 Birchmount Road
Scarborough, Ontario MIK 5G4, Canada

16   15   14   13   12   11   10   9   8   7   6   5   4   3   2   1

Library of Congress Cataloging-in-Publication Data
Physical chemistry of food processes: fundamental aspects/edited by
  Ion C. Baianu.
    p. cm.
  ''An AVI book.''
  Includes bibliographical references and index.
  ISBN 0-442-00580-6 (v. 1)
    1. Food—Composition. 2. Food industry and trade—Quality
control. I. Baianu, Ion C.
TP372.5 P49   1992
664—dc20   91-46028

Contents

PART II: APPLICATIONS TO FOOD PROCESSES

Preface

Recent trends in food science and technology are toward engineering applications based on improvements of our understanding of molecular processes and interactions between food components; hence the rapid development of physical chemistry applications to the study of food processes. The use of state-of-the-art techniques and the precise tools they offer for the detailed investigation of complex, biological and food systems, require skills similar to those utilized in biotechnology, genetic engineering, biophysical chemistry, and molecular medicine. The stage of rapid development that food engineering is now experiencing creates a unique opportunity for formulating new foods with improved stability and specially selected compositions for superior nutritional, dieting, and medical qualities.

Student training in these new and rapidly developing areas of food science and biotechnology is becoming increasingly more difficult and demanding, especially because of the lack of textbooks that are not outdated or simply inadequate. In specialized areas, several texts are now being prepared or printed that do not, however, cover the basic physical and chemical concepts necessary for problem solving and a thorough understanding of the processes that occur in foods. *Physical Chemistry of Food Processes* is an attempt to address the current needs of food science students, food scientists, and biotechnologists for a basic treatment of food processes that relates structure to functionality and directly applies the principles of physical chemistry to foods. Such applications need to be related to experiments and to the use of various physical and chemical techniques for investigating molecular structure and functionality. I planned a two-volume book to cover both the fundamental aspects and the more advanced techniques and applications, with the material being organized essentially in three parts that cover (1) principles, basic concepts, and molecular structures; (2) techniques; and (3) processes, interactions, and applications. The two volumes are intended to be used not only by food scientists and technologists, but also by biochemists, physical biochemists, biologists, and applied chemists (e.g., polymer chemists and chemical engineers). I have tried to make this book useful to both undergraduate and graduate students, as well as to instructors, with a wide range of backgrounds. Depending on the level of the course, the material in the book can be covered in one or two semesters, with one semester being sufficient for Volume I, the topics in which are briefly presented as follows.

A large number of foods occur in the form of food dispersions; an overview of such complex systems and their physical and chemical properties is presented in Chapter 1. The material is in outline form so it can be readily assembled for presentation in the classroom and to allow the student to quickly find the relevant section for reference use. Both microscopic and macroscopic bulk properties of food dispersions/"colloids" are discussed. Special emphasis is placed on rheological aspects or flow properties of food dispersions. Several types of food processing involve heat exchanges that are governed by thermodynamic laws and principles. Therefore, improvements of heat-processing methodology and equipment require an understanding of both thermophysics and chemical thermodynamics. A simplified presentation of basic thermodynamic concepts, laws, and principles is given in Chapter 2. The concepts of entropy, free energy, and reversible and irreversible processes are introduced, and specific examples are given to illustrate such concepts. Chapter 3 introduces in a simplified form basic quantum mechanical concepts that are essential to understanding chemical bonding and molecular structures. Chemical structures of major food components and biomolecules are presented in Chapter 4, often with supporting experimental data, such as their X-ray diffraction patterns from crystals. (A section on food enzymes was deliberately omitted because of the complexity of the subject and the availability of excellent recent textbooks on food enzymes and enzymatic reactions in foods.) Structural and spectroscopic techniques are briefly introduced in Chapter 5. Modern, powerful techniques such as nuclear magnetic resonance and X-ray/neutron diffraction are emphasized. Several examples of uses of electron microscopy, electron spin resonance, and calorimetric techniques to food analyses and the study of food components are also presented.

The second part of Volume I is concerned with food processes and applications. Molecular interactions that are involved in food processes are discussed in Chapter 6, together with nonideal behavior in concentrated dispersions and solutions, hydration of foods and food components, water sorption, and salt binding in foods. These important topics are rarely covered in terms of modern concepts and theories in food textbooks as done in this chapter. Further reading on the relevant modern models and theories is strongly suggested, and most of the best references are included for this purpose. Specific food processes and applications are discussed in the last three chapters, covering topics such as food extrusion (Chap. 9), rheology of cheese (Chap. 8), and modeling of processes in the pilot plant (Chap. 7).

The second volume provides up-to-date information that is not currently available in any standard physical chemistry textbook and was made possible only by the generous participation of several contributors who are actively involved in developing this field through their research and publications. We feel, therefore, that the contents are rather unique and provide both university students and researchers with a thorough and current presentation of exciting developments in the field. Specific topics covered include applications of scattering and spectroscopic techniques to food proteins and food systems; lipid structures and biotechnology applications; and applications to food processes.

Acknowledgments

During the preparation of the book I received support from various sources. Among the most important contributions to Volume I that are here gratefully acknowledged are those made by the following people.

Professor Toshiro Nishida provided me with his food chemistry course notes as a guide to the protein section in Chapter 4. Professor Patricia V. Johnston kindly provided me with her food chemistry course notes as a guide to the carbohydrate and lipid sections in Chapter 4. Dr. Eiichi Ozu often helped me with many of the computer illustrations. Thanks are also due to Mrs. Wynemia Lindsay who patiently typed a major portion of the manuscript.

Contributors

Mr. Michael F. Kozempel
Dr. Peggy Tomasula
Eastern Regional Research Center
U.S. Department of Agriculture
600 East Mermaid Lane
Philadelphia, PA 19118

Dr. Michael H. Tunick
Dr. Edward J. Nolan
Eastern Regional Research Center
U.S. Department of Agriculture
600 East Mermaid Lane
Philadelphia, PA 19118

PHYSICAL CHEMISTRY OF FOOD PROCESSES

VOLUME I

Fundamental Aspects

I

Principles, Structures, and Techniques

1

Introduction: From Principles to Applications

From Chaos, Eons upon Eons of Time ago, The Universe was born,
and within It, the Wonderful Evening Star, Hyperion, shining bright,
aloof and forlorn

Mihai Eminescu: ''The Evening Star'' (Luceafarul)

Teaching physical chemistry applications to life sciences or to food science students is a challenging task for several reasons. Living cells, organisms, and foods are dynamic, complex systems with nonuniform distributions of small molecules and biopolymers. Apart from the complications of heterogeneity, compartmentation, and numerous components, there is also the complex nature of multiple molecular interactions (nonideality) and nonequilibrium processes occurring in such systems.

Perhaps one of the easier ways of approaching the physical chemistry of food processes is to first consider simpler processes and then to progressively add new features that correspond to real aspects of food processes. The application of powerful concepts and modern techniques (which are part of modern physical chemistry) to food systems allows one to obtain meaningful results and make practical, useful predictions.

THE APPROACH

We start by considering the principles that govern molecular structures and processes. To progress beyond this point toward applications one needs to consider larger, complex structures and specific cases in which these principles apply.

A *food process* is considered to be a *change*, or *sequence of changes*, occurring in a food system, and can be both physical and chemical. Let us consider three examples of complex food processes: (1) corn wet-milling; (2) extrusion of foods; and (3) microwave cooking. The second example will be considered in much greater detail in Chapter 9.

3

A greatly simplified diagram of corn wet-milling is presented in Figure 1-1. At the beginning of this process, corn kernels are "steeped" in a solution containing bisulfite. The lowering of pH caused by the SO_2 dissolved in water, or by bisulfite, inhibits the growth of most bacteria, except lactobacilli. The chemical reactions of sodium bisulfite with the corn storage proteins, such as

$$NaHSO_3 + R-S-S\text{-protein} \rightleftharpoons RSSO_3^- + HS\text{-protein} + Na^+ \quad (1\text{-}1)$$

result in an easier separation of the corn starch from the corn proteins that trap starch granules. The lactic acid produced by the lactobacilli also acts to improve the corn starch separation from the corn proteins. Apart from such chemical processes there are a number of physical processes that occur during steeping. Among them are the increased hydration of starch and proteins, as well as the solubilization of numerous compounds present in the corn kernels (e.g., ions, low molecular weight carbohydrates, and certain amino acids). Following the separation of oil and starch from proteins, the processing is concerned mainly with corn starch. Enzymatic processing of corn starch yields high-fructose corn syrups, or maltodextrins. Various chemical modifications of the corn starch

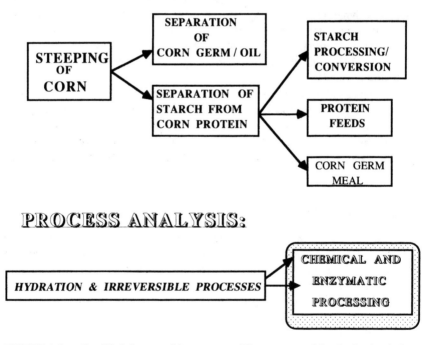

FIGURE 1-1.   Simplified diagram of the corn wet-milling process and the physicochemical processes involved.

itself may also be employed to obtain a variety of "bulking agents" for foods, gelling starches, amylopectins, or other carbohydrates for specialized applications. In this complex food process one can readily identify several physical and chemical processes, such as

- Diffusion of water and hydronium ions into the corn kernel
- Hydration of proteins and starch
- Sedimentation of starch
- Chemical reactions between sodium bisulfite and corn storage proteins
- Hydrolysis of starch, or selected chemical modifications
- Solubilization of water-soluble components from the corn kernel
- Enzyme reactions involving the hydrated corn starch as a substrate
- Separation of corn oil

The preceding list is far from being complete, and the details of the starch–protein separation process are yet to be elucidated. However, the corn wet-milling example does include several physicochemical processes that are also encountered in many other important food processes. Among these physico-chemical processes are water diffusion, sedimentation, hydration of proteins and starch, and enzyme reactions. Several of these processes will be considered in some detail in Chapters 6 to 9.

The second example selected here involves the extrusion processing of foods as schematically illustrated in Figure 1-2. Following the mixing of a solid powder of a food material (such as flour) with water, a food "dough" is formed. Its flow properties determine to a large extent, but not completely, the process occurring in the extruder. Among the additional processes occurring are "homogenization," phase transitions, and chemical reactions caused by heating in the extruder barrel, especially in the metering section (Fig. 1-2; see also Ch. 9). Diffusion processes, molecular distributions, distribution of residence times of food particles in the extruder, and expansion of the dough at the die as the steam evaporates rapidly, are all processes that need be considered in the operation of an extruder. Still, there are other properties and processes that need to be considered, among which are sheer thinning, inhomogeneous distribution

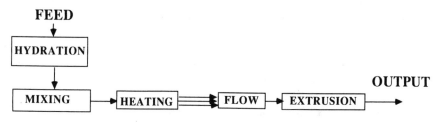

FIGURE 1-2.   Simplified diagram of the food extrusion process.

of water in the food dough, heat transfer and heat capacity, residual strains, viscoelastic properties at the higher concentrations, thoroughness of mixing, leakage flows, and barrel "end effects." A more detailed presentation of the extrusion of foods is deferred to Chapter 9, where the physical and chemical processes involved in extrusion are discussed in relation to processing of starch and food doughs. Most of the analysis of the extrusion process is, however, centered around the flow processes, or the rheology of food doughs in the extruder. From this example of the complex extrusion process one sees, as in the preceding example of Figure 1-1, that heat transfer and thermodynamics are involved in an essential manner. Furthermore, *rheology* plays a central role, in addition to thermodynamics.

The third example, that of microwave cooking of foods will not require a diagram since the process, at least in its practical form, is undoubtedly familiar to the reader. However, the principles underlying this process may not be as widely known or understood. The food contains water molecules, whose electric dipoles absorb the microwave energy. This energy is then irradiated as heat by water molecules. Water hydrating the proteins, carbohydrates, and lipids present in the food absorbs microwaves differently from the "free" water that is just trapped in the food. Last, but not least, the water hydrating the food has a number of *discrete energy levels*, and the microwave energy induces transitions between these levels in a highly specific or selective manner. The proper understanding of this absorption process requires both a study of dielectric relaxation and quantum mechanics. The heat transfer, phase changes, and chemical reactions caused by microwave cooking are, again, subject to thermodynamic laws. Therefore, at least certain aspects of thermodynamics, dielectric relaxation, and quantum mechanics need to be understood before one can adequately treat microwave processing of foods from a physical and chemical standpoint. One could say, of course, that from a technological, or engineering, viewpoint such careful and detailed considerations may not be necessary, or that they would be "too difficult" to formulate. This technological viewpoint can be considered acceptable for simple, or limited, applications of microwave processing, but it is quite unacceptable from the standpoint of efficiently designing microwave ovens, or for engineering new microwaveable foods with improved texture, taste, quality, etc. This volume would hopefully remove the objection of a subject being "too difficult" for the patient student.

Emulsions and lipids are often encountered in foods. A lipid molecule, such as a phospholipid, has a specific *molecular structure* (see Fig. 4-21) that is responsible for its behavior in the presence of other molecules, such as water and proteins. One end of the molecule is a *dipole* (e.g., it contains a pair of two separated charges of opposite sign), whereas the rest of the molecule is a rather long alkyl chain that is nonpolar, and therefore quite different from the dipole end of the molecule. Such *amphipatic* properties of the phospholipids lead to the formation of lipid bilayers, membranes, micelles, and several phases with complex behavior and interesting surface, or barrier properties. These properties, as well as *lipid oxidation*, are determined by the molecular, or chemical, structure of lipids together with their molecular *interactions* with other molecules, such as water and molecular oxygen. The formation of emulsions and various lipid phases, as well as their molecular interactions, are best understood in terms of *molecular structure and thermodynamics*. The stability of a number of foods depends critically on the behavior of lipids in foods, including chemical reactions, such as lipid oxidation. Similar considerations apply to two other major food components, the carbohydrates and proteins. A knowledge of their chemical structures is a prerequisite for understanding processes occurring in foods. Furthermore, thermodynamics needs to be combined with a knowledge of chemical structures in order for us to be able to *predict*, or *determine*, the probable occurrence of a certain process either in foods or in other systems.

Modern physical and chemical techniques are *powerful tools* for investigating molecular structures and processes in a wide range of systems, including foods. Such techniques, employed in conjunction with the appropriate physical and chemical theories, provide detailed answers to most of the problems raised by technology or engineering. Therefore, a knowledge of the capabilities, uses, and limitations of these techniques provides the student with a deeper understanding of molecular structures and processes, as well as with the means for finding experimental answers to practical or processing questions. A number of these techniques are presented briefly and in a simplified manner in Chapter 5. A substantially more detailed and in-depth presentation of such techniques is planned for the second volume of this book. Table 1-1 illustrates the potential applications of one technique, nuclear magnetic resonance (NMR), for investigating various aspects of foods and food processes. This table is far from being complete or comprehensive (additional details are given in Table 1-2). Fur-

TABLE 1-1   Potential and Current Applications of NMR in Foods

Applications	NMR Technique	Example
Analytical	High-resolution NMR	Percent of unsaturated fatty acids in a mix
Quality control	Pulsed/low-field NMR	Determination of water in butter
Wheat flour milling	NMR imaging (MRI)	Conditioning of wheat grains
Research and development	Two-dimensional NMR	Noninvasive monitoring of bacterial growth

Note: See also Table 1-2.

TABLE 1-2   Some Applications of Physical and Chemical Techniques to Foods

Technique	Applications
FT-IR/NIR	Protein, moisture and oil determination in foods; structure determination
ESR	Determination of iron cations in soy flour; monitoring of microencapsulated materials; fiber orientation in extrudates
NMR	On-line and off-line quality control; MRI imaging of fat in meat; chemical composition analysis; formulation of new food products; determination of heat processing effects in foods; determination of structure–functionality relationships in foods; determination of food component hydration and stability; kinetics of aggregation and gelling behaviors; monitoring of phase transitions; diffusion and molecular dynamics in foods; effects of ions and sugars
X-ray diffraction and scattering	Determination of structure and hydration; starch gelatinization and formation of lipid complexes (see also Ch. 9)

thermore, other techniques, such as differential scanning calorimetry (DSC), are being increasingly used in foods, covering a wide range of different applications. Some of the widely used techniques and their uses are tentatively indicated in Table 1-2; however, no claim to completeness is made.

A suggestive summary of the physical and chemical theories involved in the treatment of various food processes is presented in Table 1-3. Again, we do not claim that the listing is complete, but the table clearly indicates the usefulness of thermodynamics (and its several branches) and rheology in the study of food processes. Furthermore, the physical chemistry of foods is a fast-expanding and developing field, in which major advances were made during the 1980s.

TABLE 1-3   Applications of Physical and Chemical Principles, Models, and Concepts to Foods and Food Processes

Theory	Food Application
Thermodynamics	
Thermophysics	Heat processing of foods; DSC and DTA analyses; design of food processors, ovens, etc.
Chemical Thermodynamics	Biochemical processes in foods; biotechnology applications; design of bioreactors/fermenters
Quantum mechanics/ Electromagnetism	Microwave processing; prediction of molecular properties of food components; spectral analysis of food materials
Hydrodynamics/Rheology	Functional properties of foods; extrusion of foods; flow control in food processing plants

To sum up the approach proposed so far, one should begin by considering *basic principles*, give *examples*, and then add *structural information* and observations made by powerful, modern *techniques*. As a next step, one would consider the relevant *molecular interactions* and *processes*, and finally *integrate* the various *physical* and *chemical processes* into the overall treatment, or model, of the food process in order to make quantitative estimates and predictions.

DISPERSED PHASES IN FOODS

In this section we present an overview of food dispersions; that is, classification, dispersion stability, other major properties, and flow. Because foods are complex mixtures of several components and phases that interact in specific ways, it is necessary to introduce a classification of foods in terms of such phases/components that would help us understand them and compare their properties. A brief overview of the major properties of foods and their classification will serve as a guide to map out specific areas in foods and to develop specific applications in detail. The best guides are short, well-organized, and contain specific examples of illustrations of the main points. The concepts defined in this section are amplified later in the context of specific applications. For example, rheological concepts introduced here are also encountered in the contexts of cheese rheology in Chapter 8 and extrusion rheology in Chapter 9. The following presentation is therefore set out in outline form to give the reader a bird's-eye view of the wide range of topics.

    I. Food systems and food dispersions: Classification
      A. Food systems: Classification
        1. Edible tissues
          a) Fruits
          b) Vegetables
          c) Meat
        2. Food dispersions
          a) Milk
          b) Tomato juice
          c) Mashed potato
          d) Meat emulsion

Definition. A *food dispersion* is a continuous phase with one or more discontinuous (*dispersed*) phases. Table 1-4 lists the dispersions with only *two* phases; their properties are briefly presented in the remainder of this section.

TABLE 1-4   Dispersions with Two Phases

Dispersed Phase	Continuous Phase	Name	Example
S	L	Sol	Skim milk
L	L	Emulsion	Salad dressing
G	L	Foam	Meringue
G	S	Solid foam	Foam candy, cake
S	G	Solid aerosol	Smoke
L	S	Gel*	Gelatin

Notes: S = solid; L = liquid; G = gas.

*An SL dispersion with *both* phases being *continuous* is called a *gel* and consists of a three-dimensional network of the polymer, with the liquid phase filling the remaining space; also, it can be considered as a dispersed liquid in a solid, *continuous* phase (the reverse of a sol).

B. Dispersion stability
   1. Size distribution (Table 1-5 illustrates the effect of particle size on viscosity for spherocolloids)
   2. Interfacial tension
   3. Density
   4. Viscosity
   5. Electrical repulsion (The rate of motion of particles, dispersion stability, and the other factors are considered next)
      a) Fine particles give more stable dispersions.
      b) Interfacial/surface tension (The surface tension is caused by the surface molecules being in a different *environment*; too high a surface tension decreases the dispersion stability (see also emulsion stability))
         (1) H-bonding and Van der Waals forces produce *net attractive forces* toward the interior of the liquid
         (2) The interfacial area A is minimized

TABLE 1-5   Effect of Particle Size for Spherocolloids

Viscosity of Glycogens of Different Molecular Weight (M_w) in 0.1 M $CaCl_2$

Mol. Wt.	$[\eta]$	k
1,530,000	0.078	0.12
450,000	0.085	0.13
200,000	0.083	0.12
110,000	0.081	0.12
37,000	0.082	0.12

(3) The interfacial surface tension, σ, is measured in dynes/cm

(4) As σ increases, A decreases

(5) The effects of solutes in aqueous solutions

    (a) Inorganic salts and polar molecules increase σ

    (b) Surfactants decrease σ

c) The presence of charges tends to stabilize a dispersion

d) The more viscous dispersions are generally more stable

C. Optical properties

  1. Light is scattered by colloidal* particles, especially when they are not highly solvated

Definition. The phenomenon of light scattering by colloidal particles is called the *Tyndall effect*. The scattered intensity is $I_R \simeq c/d$, when the particles have diameters, $d \geq 1$ μm, with c being the concentration of the suspension. An example of suspension is provided by the casein micelles in milk or in water.

D. Bulk properties of dispersions

  1. Slow diffusion of the dispersed phase

  2. Large surface-to-volume ratio

  3. Brownian movement

  4. Tyndall phenomenon/light scattering

  5. High interfacial energy

  6. High adsorptive capacity

  7. Sizes between 10 and 5000 Å, with the latter being close to the resolving power of the light microscope

E. Electrical properties

  1. Dispersed phase

    a) Charges (These result from either the ionization of an ionizable group or absorption of charged ions from solution due to the presence of dipoles in the macromolecule).

    b) Dipole moments

  2) Examples of charges in food dispersions

    a) Sulphate ions on polysaccharides (The sulphate group behaves as a *strong electrolyte*, and remains ionized throughout the pH range, except in the presence of strong cations)

    b) Carrageenan

    c) COO$^-$ *alginate*

    d) SH in β-lactoglobulin

    e) The tyrosine hydroxyl group

---

*In Greek "colloidon" means glue.

      f) PO_4^{2-} in phosphoproteins

      g) The amino group of proteins

      h) Hexosamine in polysaccharides

      i) Guanidinium and histidine groups in proteins (e.g., myosin)

      j) *Amylose–iodine* complex, where I_2 and I^- are involved

F. Stabilizing agents

  1. Emulsifiers (i.e., surface-active agents, either ionic or nonionic)

    a) Ionic

      (1) FA salts

        (a) Soaps

      (2) Na stearoyl-2-lactylate

      (3) Phospholipids

      (4) Proteins

    b) Nonionic

      (1) Mono- and di-glycerides

      (2) Polyglycerol esters of FAs

      (3) Propylene glycol esters of FAs

      (4) Sorbitan FA esters

      (5) Polyoxyethylene sorbitan FA esters

        (a) Polysorbate finely divided solids

        (b) Coagulated protein

        (c) Ground spices

  2. Effects of emulsifiers

    a) To increase the viscosity, η, of the water phase

    b) To form interfacial films on oil droplets

II. Sols

A. Terminology

  1. Molecular dispersion ($d < 1$ nm, a true solution)

  2. Colloidal dispersion (with particle sizes, d, less than 0.5 μm)

    a) Macromolecules

    b) Micelles

      (1) Milk

  3. Course dispersion (d larger than 0.5 μm)

    a) Macroparticles (These are less stable and have higher plasticity)

      (1) Peanut butter

      (2) Applesauce

      (3) Ketchup

  4. Lyophobic sols (The contrasting properties of lyophobic and lyophilic sols are compared in Table 1-6; a *lyophobic* colloid being one with particles having very little affinity for the dispersing liquid)

B. Rheology of sols

  1. Viscosity

TABLE 1-6   Properties of Lyophobic Versus Lyophilic Sols

Redispersibility	Irreversible	Reversible
Resistance to electrolyte	Only low	High
Concentration of the dispersed phase	Only low	High possible
Viscosity	About the same as the dispersing medium	Higher than that of the dispersing medium

Definition. Viscosity is a measure of the *resistance* of a fluid *to flow* or shear. The viscosity of a sol is highly dependent on the surface characteristics of the particles in the sol.

Definition. A *flow curve* is a plot of shear stress against the rate of flow or shear rate.

      a) τ = shear stress, the force applied to the top plane in a parallel-plate configuration, with flow between the plates (Fig. 1-3)
      b) $\dot{\gamma}$ = shear rate, the gradient of fluid velocity = dv/dt
  2. Measurements of viscosity

Note. Only *apparent viscosity* is measured for non-Newtonian food systems. The Brookfield viscometer is commonly used for this determination. For a given shear stress, the shear rate is determined. Time must be specified in order to get a reproducible reading, because structural breakdown is common. In addition, temperature must also be specified. Because this measurement is empirical, comparison of viscosities can only be made when measured under *identical* conditions.

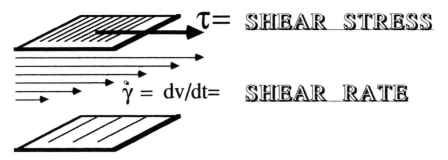

τ = SHEAR STRESS

$\dot{\gamma}$ = dv/dt= SHEAR RATE

FIGURE 1-3.   Definition of viscosity for Newtonian flow between two parallel plates.

III. Gels
    A. Gels are dispersions with *two continuous phases* (Table 1-4)
        1. Thermo-reversible
           a) Amylopectin gels
        2. Thermo-irreversible
           a) Amylose gels
           b) Cottage cheese
           c) Curd
    B. Syneresis over time
        1. Retrogradation of a starch gel
           a) Puddings
           b) Frozen foods
    C. Thixotropy and rheodestructive gels

Definition. In some sols or gels the shear rate increases over time under constant shear stress because the viscosity is changing as a result of structural breakdown. If the viscosity returns to the initial value after a rest period, the sol, or gel, is called *thixotropic*; if the viscosity remains low, the sol is called *rheodestructive*.

IV. Emulsions
    A. Breakdown of emulsions
        1. Creaming (the formation of two emulsion layers)
        2. Flocculence (the agglomeration of droplets)
        3. Coalescence (the irreversible union of small droplets, leading to the separation of two phases)

Definition. An *emulsion* is an immiscible liquid dispersed in a continuous liquid phase (e.g., water). An emulsion is generally unstable without the presence of emulsifying agents that lower the interfacial tension.

Definition. *Phase separation* from emulsion can often be described by the Stokes equation:

$$v = 2r^2 \cdot g(d_1 - d_2)/9\eta, \qquad (1\text{-}2)$$

where
      v = velocity of separation
      r = particle "radius"
      η = viscosity
      g = gravitational acceleration
  d_1, d_2 = densities of the two phases (Einstein, 1911).

V. Foams
   A. Foams are formed by a dispersed gas in a liquid or solid phase
      1. Whipped cream
      2. Ice cream
      3. Cake
      4. Bread
      5. Marshmallow
      6. Meringue
      7. The head on beer
   B. Liquid foams are thermodynamically unstable (see Chap. 2) due to the free-energy decrease when the foam collapses
      1. The loss of liquid in the bubble walls is a major reason for the collapse, and occurs because of
         a) Gravity
         b) Evaporation
         c) Deformation stress due to gas diffusion to large bubbles
      2. Foam stabilization
         a) Increases viscosity, η (e.g., the effect of gums and proteins)
         b) Interface stabilization (caused, for example, by proteins)
      3. Antifoaming agents, such as dimethyl polysiloxanes, displace wall liquid, thus thinning the foam bubble walls
VI. Viscosity of colloidal systems
   A. Most foods are non-Newtonian
      1. Pseudoplastic
         a) Milk
         b) Dilute hydrophilic sols, especially gums
      2. Dilatant
         a) A 35% starch slurry
      3. Non-Bingham plastic
         a) Ketchup
         b) Applesauce
   B. The relative viscosity

$$\eta_{rel} = \frac{t_1 \rho_1}{t_0 \rho_0} = \frac{\eta_1}{\eta_0}$$

where t_1 is the time required for a given volume of the colloidal lyosol to flow through a capillary, and ρ_1 and ρ_0 are the densities of the lyosol and the medium, respectively.
   C. The specific viscosity

$$\eta_{sp} = \eta_{rel}$$

D. The reduced viscosity

$$[\eta] = \eta_{sp}/c$$

where c is the number of grams of dispersed material per 100 cm^3.

E. The viscosity of spherocolloids
   1. Dependence on *concentration* very small (Table 1-5)

Definition. At low concentrations, with negligible hydration, the *specific viscosity* of spherocolloids is a *linear* function of the concentration:

$$\eta_{sp} = 0.025 \cdot c/\rho = 0.025\phi \qquad [\eta] = (1/4)\phi/c$$

ρ = the density of the dispersed phase

ϕ = the volume percent occupied by the sphere

As the concentration increases, interactions will occur; therefore,

$$\eta = [\eta] + k^1 \cdot [\eta]^2 \cdot c, \tag{1-3}$$

where $[\eta]$ is the *intrinsic* vicosity, and k^1 is the "Huggins" constant with a value close to 2.0.

   2. Dependence on *particle size* or mass (see Table 1-5)
F. Intrinsic viscosity of macromolecules

$$\eta_{sp}/c = A + Bc + \cdots$$
$$[\eta] = \lim_{c \to 0} (\eta_{sp}/c) = A \tag{1-4}$$

Definition. The *intrinsic viscosity* of a macromolecule depends on its shape (through A) and its *specific volume* (Morawetz, 1965).

G. Viscosity for a dilute solution of rigid spheres

$$\eta_{rel} = \eta/\eta_o = 1 + \phi \cdot c + 1 = 2.5c$$

or

$$\eta_{sp} = (\eta/\eta_0) - 1 = 2.5c$$

Definition: The *specific viscosity* of a dilute solution of *rigid spheres* is a linear function of the concentration

$$[\eta] = 2.5 \cdot c$$

(The value of 2.5 for ϕ is a minimum. Any deviation from sphericity leads to a larger value; Flory, 1953.)

H. Viscosity for a dilute solution of hydrated spheres

$$\eta_{sp} = 2.5[c + (w/\rho)c]$$

where
w = grams of solvent bound per gram of anhydrous solute
ρ = density of solvent
$[\eta] = 2.5 [1 + w/\rho] \cdot c$

I. Dependence of viscosity upon concentration

$$\eta_{sp}/c = [\eta] + k[\eta]^2 c + \cdots \qquad (1\text{-}5)$$

where k is the "Huggins" constant, and $\tau/\dot{\gamma} = \eta$, is the viscosity, measured in *poise* (P):

$$1P \equiv 1 \text{ dyne} \cdot cm^{-2} \cdot s^{-1}$$

1. Proportionality between shear stress and shear rate

Definition. When the shear stress and shear rate are *proportional*, there is *Newtonian flow* (line (1) in Fig. 1-4).

2. Dispersed phase hydration

Hydration of the dispersed phase increases the apparent value of the constant ($k \rightarrow 0.025$), since the hydration increases the effective hydrodynamic volume of the particles.

3. Effect of particle size for spherocolloids
   a) For spherical particles of the same density, the reduced viscosity is *independent* of particle size, except for extremely small particles ($d < 100$ Å) (see Table 1-5)
4. Effect of charge (Fig. 1-5)

SHEAR STRESS

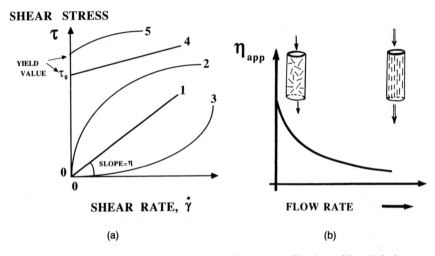

(a)   (b)

FIGURE 1-4.   (a) Major types of flow behavior (flow curves). The slope of line (1) is the *viscosity* for Newtonian flow, whereas the slope of line (4) is the *Bingham viscosity*. The *yield value*, τ_0, is characteristic of plastic flows (either Bingham or non-Bingham). (b) Schematic representation of the orientation of large molecules or particles that occurs at the higher flow rates for non-Newtonian flows, such as pseudoplastic ones (curve (2) in Fig. 1-4a).

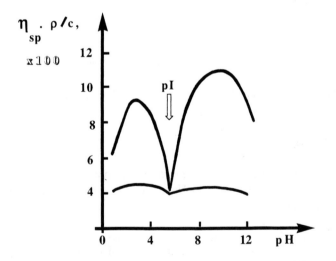

FIGURE 1-5.   Charge dependence (pH-titration) of intrinsic viscosity for 0.2% egg albumin solutions with, or without, added salt ions (Na^+ and Cl^-). The pH was varied by the addition of appropriate volumes of 0.02 M NaOH or 0.01 M HCl solutions. The top curve is without salt, whereas the lower curve is with 0.1 M NaCl added.

A charged system will be more viscous than an uncharged system; for example, 0.2% egg albumin in an aqueous solution, with or without salt.

J. Viscosity of food sols
   1. Newtonian type

Definition. The rate of shear (gradient in fluid vicosity, $\dot{\gamma}$) and shear stress, τ, are linearly related. The proportionality constant is the *viscosity* (η):

$$\tau = \eta \cdot \dot{\gamma} \qquad (1\text{-}6)$$

   2. Non-Newtonian type

Definition. The plot of shear stress vs. shear rate deviates from the ideal, Newtonian behavior (Fig. 1-4). There are five types of flow behavior shown in Figure 1-4:

- Newtonian, η = slope
- Pseudoplastic
- Dilatant
- Bingham plastic
- Non-Bingham plastic

The last four types of flow are non-Newtonian.

   3. Plastic systems
      a) Many food systems behave like "soft" plastics, and are therefore called *plastic systems*
         (1) Jellies
         (2) Gums
         (3) Ketchup
         (4) Applesauce
         (5) Fudge
      b) The property *yield value* is the shear stress needed to initiate the flow
      c) Bingham and non-Bingham plastic flows
   4. Pseudoplastic flow (The apparent viscosity decreases with increasing rate of shear)
      a) Dilute hydrophilic sols
      b) Milk
      c) Gum sols
   5. Dilatant flow (The apparent viscosity increases with increasing shear rate)
      a) A 35% starch slurry

      6. Thixotropy
        a) Concentrated juices
        b) Jellies

Definition. *Thixotropy* is a *reversible gel–sol transformation* in which a gel is converted to a sol by agitation or shear stress; upon removal of the disturbing force, the consistency increases and the gel is reformed.

  L. The viscosity of linear colloids
     1. Reduced viscosities

The reduced viscosities of linear colloids are much higher than those of spherocolloids, and are dependent on the following variables

- Concentration (Fig. 1-6)
- Particle size (Fig. 1-6)
- Shape
- Hydration
- Charge (Fig. 1-5)
- Shear stress

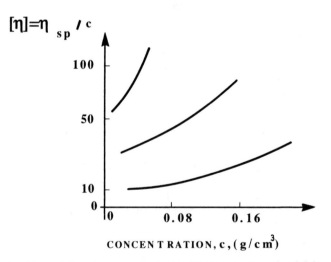

FIGURE 1-6.   The variation of reduced viscosities with concentration and molecular weight for dispersions of linear biopolymers. The top curve is for biopolymers of about 1 million M_w, the middle curve is for M_w on the order of 200,000, and the bottom curve is for linear biopolymers of M_w around 60,000.

2. Shear stress and orientation

As the rate of flow increases, the apparent viscosity of linear colloids decreases due to the *orientation* of the linear molecules (Figs. 1-4*a* and *b*).

3. Moderately concentrated linear colloids

The viscosities of moderately concentrated linear colloids are non-Newtonian, because the viscosities are dependent upon the flow rate or shear stress.

4. Yield values and apparent viscosities

Note. There is *no correlation* between yield values and apparent viscosities. For example, mayonnaise has a moderate yield value but relatively low apparent viscosity, whereas sterile condensed milk (before gelling) has a zero yield value but a relatively high apparent viscosity.

Summary

Viscosity

Definition. *Viscosity* is a measure of the *resistance* of fluids *to flow*.

The absolute unit of viscosity is the *poise*, which is the force of 1 dyne $\cdot$ cm^{-2} creating a 1-$cm^2 \cdot sec^{-1}$ velocity gradient across the distance of 1 cm. Water at 20°C has a viscosity of 1.0 centipoise (0.01P).

The *relative viscosity* is the ratio of a solution viscosity (η) to the solvent viscosity (η_0),

$$\eta_{rel} = \frac{\eta}{\eta_0} = \frac{t \cdot \rho}{t_0 \rho_0}$$

where t is the time required for a given volume of the solution to flow through a capillary, t_0 is the time required for the same volume of the solvent to flow through the capillary, ρ and ρ_0 are the density of the solution and the solvent, respectively, and η and η_0 are the viscosity of the solution and the solvent, respectively.

The *specific* viscosity

$$\eta_{sp} = \eta_{rel} - 1.0$$

The *reduced* viscosity

$$\eta_{rel} = \frac{\eta_{sp}}{c}$$

where c is the concentration in grams per milliliter.
Intrinsic viscosity

$$[\eta] = \lim_{c \to 0} (\eta_{sp}/c)$$

The general equation for the viscosity:

$$\eta_{rel} = \eta/\eta_0 = 1 + A\phi + B\phi^2 + \cdots \qquad (1\text{-}7)$$

where ϕ is the volume fraction of particles, and $A, B, \cdots$ are coefficients that depend on the nature of the solute (*virial* coefficients).

$$\eta_{sp} = \eta_{rel} - 1 = \frac{\eta - \eta_0}{\eta_0} = A\phi + B\phi^2 + \cdots$$

$$= Ac + Bc^2 + \cdots \qquad (1\text{-}8)$$

where c is concentration and ϕ is the specific volume of solute (Flory, 1953).

Theory of Emulsion Formation and Stability

The formation of a *stable emulsion* requires that the interfacial tension be sufficiently *low* to allow for a large increase in surface when a reasonable amount of work, or increase in energy, is applied. Also, the coalescence of the globules must be prevented, or retarded, by an energy barrier, either electrical or mechanical, that *exceeds* the kinetic energy of the globules (Becher, 1935).

When an emulsifying agent is present, the interface actually has two faces (i.e., one on each side of the interfacial film). Thus the dispersed phase will be the phase with the higher interfacial tension (with the film as the emulsifying agent), such that the energy requirements for formation are minimized (Fig. 1-7).

In the adsorption "theory" of emulsion stability, the emulsions are stabilized by a monomolecular film of emulsifier with the polar groups oriented toward the aqueous phase and the nonpolar groups oriented toward the oil phase. Even after four years, globular size in octane emulsions stabilized by 0.1-M Na-oleate solutions remain unchanged. On the other hand, octane emulsions in 0.005-M Na-oleate solutions are much less stable (see Fig. 1-8). Calculations indicate that the emulsion became stable at a mean interfacial area per molecule of soap between $\simeq 27$ and 45 Å^2.

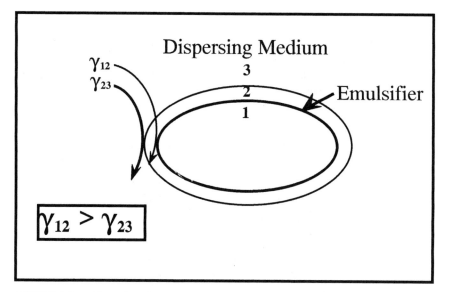

FIGURE 1-7.   Illustration of the action of an emulsifier (2) between two dispersed phases (1) and (3). The condition for emulsion formation is that the appropriate surface tension coefficients satisfy the relation $\gamma_{12} > \gamma_{23}$.

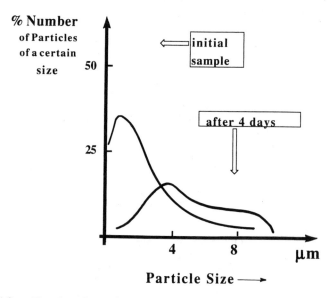

FIGURE 1-8.   Time dependence of coalescence in octane emulsions with 0.005-M Na-oleate. Note the shift in the peak of the particle size distribution from ~ 1 μm at the beginning of the sequence to ~ 4 μm after four days.

The Hydrophile–Lipophile Balance

Definition. The *hydrophile–lipophile* balance (HLB) is determined by the weight-percent of that portion of the emulsifier molecule that is *hydrophilic*, divided by 5. An example is shown for polyoxyethylene sorbitan monolaureate in the following calculation:

Total Composition		
Sorbitan (1 mole)	=	164.16
Lauric acid (1 mole)	=	200.32
Ethylene oxide (20 moles)	=	881.00
Water of esterification	=	−18.02
Total Number of moles	=	1227.46

Hydrophile Portion		
Sorbitan (1 mole)	=	164.16
Ethylene oxide (20 moles)	=	881.00
Number of hydrophilic moles	=	1045.16

HLB number $= 1045.16/1227.46 \cdot 100/5 = 17$

The HLB number and dispersibility in water are correlated in Table 1-7, while the relationship between the HLB number and the type of application of the emulsifier is indicated in Table 1-8.

Table 1-7   HLB Number and Dispersibility in Water

HLB Range	Dispersibility or Solubility
1–4	No dispersibility in water
3–6	Poor dispersibility
6–8	Milky dispersion after vigorous agitation
8–10	Stable milky dispersion
10–13	Translucent-to-clear dispersion
13–20	Clear solution

TABLE 1-8   HLB Number and the Type of Application of the Emulsifier

HLB Range	Application
3–6	Water/oil emulsifier
7–9	Wetting agent
8–10	Oil/water emulsifier
3–15	Detergents
15–18	Solubilizers

FOOD DISPERSIONS: MICROSCOPIC PROPERTIES

I. Electrical properties of colloids
   A. Charge on a globule

The *charge on a globule* has its source in the inherent chemical nature of the surface of the dispersed phase that determines the sign and magnitude of the charge. There are, therefore, two sources of charge on a globule: (1) the *ionization* of ionizable groups at the surface of the globule (e.g., groups on the emulsifying agent, or the dispersed phase itself), and (2) the *adsorption of charged ions* from the dispersing medium.

   B. **Definition.** *Helmholtz* double layer
     1. *Electrical double layer* at the interface (see Fig. 1-9)
     2. The external layer of charges is "diffuse" around the particle in an *ion atmosphere*, with a preponderance of charges *opposite* in sign to that of the particle (Gouy and Stern models, the Debye–Hückel theory, and other related models)

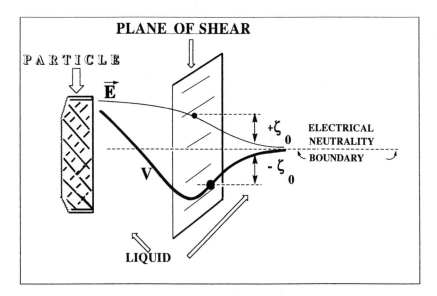

FIGURE 1-9.   The electric field and potential dependences on distance from a colloidal particle surface. The plane at arrow is the "shear plane" of the Helmholtz double layer theory (see also Figs. 1-10 and 1-11 for further details).

C. **Definition.** The *zeta potential*
    1. Due to the Helmholtz double layer, a potential gradient E_0 exists in the vicinity of the particle (see Fig. 1-10). The value of the potential at the *plane of shear* between the "*fixed*" counterion layer and the second "*diffuse*" layer is called the *zeta potential* (ζ_0) of the particle (Figs. 1-9 and 1-10).

Definition. The *amphoteric molecule* is a molecule that is capable of possessing both negative and positive charges.

D. Stability
    1. The electrical double layer surrounds a particle with a net charge
        a) There is a fixed counterion layer, adsorbed onto the particle surface (60%–85%)
        b) There is also a diffuse counterion layer, in solution (the remainder of the ions, see Fig. 1-11)
        c) In an electrical field the particle will move, leaving behind much of the "diffuse" counterion layer. The *zeta potential* is the value of the *potential at the shear boundary* (The *electrophoretic mo-*

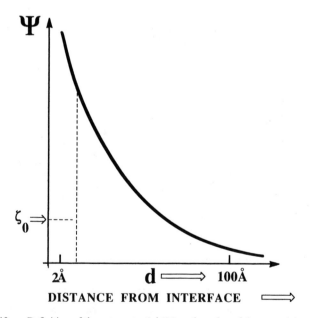

FIGURE 1-10. Definition of the *zeta potential* (ζ_0) as the value of the potential at the *shear boundary* at a distance d_0 from the particle interface with ions and water.

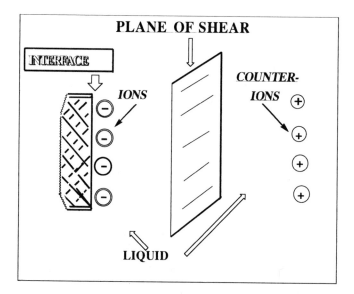

FIGURE 1-11.   Schematic representation of the plane of shear in the Helmholtz double layers of ions in water surrounding a colloidal particle.

 bility, μ, is proportional to the zeta potential) (see Figs. 1-9 to 1-11)

d) When double layers of two particles overlap, repulsion occurs between like-charge particles

e) Flocculation depends on the energy of the double layer and the extent of solvation

f) The addition of salt can reduce the zeta potential and desolvate the surface. Salting-out occurs when sufficient salt is added to cause precipitation. Much more salt is needed to precipitate a hydrophilic sol than for a hydrophobic sol.

For example, consider the effect of a neutral salt (K_2SO_4) on the solubility of carbon monoxide hemoglobin at its isoelectric pH. The ionic strength of a solution μ_z is given by $(1/2)\Sigma c_i z_i^2$, in which c is the concentration and z is the charge. At low ionic strength, the protein is *salted-in* (i.e., there is an *increase* in solubility). At high salt concentration, it is *salted-out* (i.e., there is a *decrease* in solubility), as illustrated in Figure 1-12.

II. Electrochemical concepts
   A. Isoelectric point

Definition.   The *isoelectric point* is the pH value at which the molecule has *zero* net charge (usually measured in the presence of added salt).

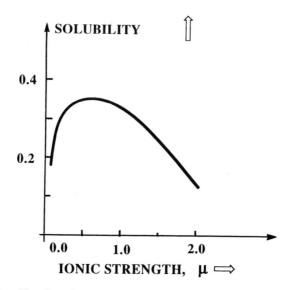

FIGURE 1-12. The effect of a neutral salt (K_2SO_4) on the solubility of carbon monoxide hemoglobin at its isoelectric pH. (The ionic strength of a solution, μ, is given by $(1/2)\Sigma\ c_i z_i^2$, in which c is the concentration and z is the charge.) At low ionic strength, the protein is *salted-in*, that is, has increased solubility. At high salt concentration, it is *salted-out*, that is, has decreased solubility, as shown in the figure.

B. Isoionic point (measured in the absence of any added salt)

Definition. The *isoionic point* is the pH value at which the number of anionic groups of the solute that have lost a proton equals the number of charged cationic groups. (An *isoionic solution* has *no* extraneous salt ions present.)

C. Electrophoresis

Definition. *Electrophoresis* is the movement of the particles of the dispersed phase relative to the dispersing medium in an *external electrical field*. The important variable is the *electrical field*, which causes the migration of charged particles, as illustrated in Figure 1-13.

$$|\mathbf{E}| = V/R = \text{Potential}/\text{distance} = \text{electrical field strength} \quad (1\text{-}9)$$

At equilibrium the friction force equals the electrostatic force.

D. Electroosmosis

Definition. *Electroosmosis* occurs when the dispersing medium is *forced* through a gel, capillary, or membrane made of the dispersed phase material.

$$|\vec{E}| = V/R$$

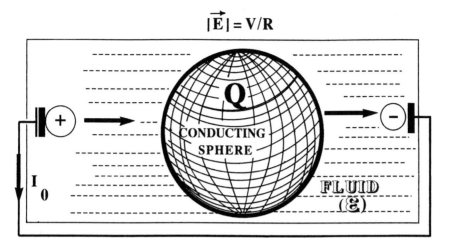

FIGURE 1-13.   Principle schematic of the electrophoresis process, showing the movement of the conducting sphere of charge Q, in a solution or gel, in a DC electrical field.

By applying a *pressure*, an electrical potential (called the *sedimentation potential*) is developed across the membrane.

E. Titration curve of protein-contributing groups
    1. Carboxyl groups
    2. Imidazole $+$ α-amino group
    3. Phenolic hydroxyl, α-amino, and SH-groups

If in the titration curve of a protein the number of groups can be roughly estimated, then a titration equation can be applied to the various types of groups present, and the titration curve can be resolved. The association for *one type of binding site* is given by

$$A_b = mK \cdot [A]e_b^{-2wA}/(1 + K[A]e_b^{-2wA}) \qquad (1\text{-}10)$$

where w is the interaction, or *"virial"* coefficient (Tanford, 1961), k is the *apparent* association rate constant, and A_b is the concentration of bound species, or ions (Scatchard, Coleman, and Shen, 1957).

F. Titration methods
    1. Potentiometric (Scatchard, Scheinberg, and Armstrong, 1950)
    2. Dialysis equilibrium
    3. Sedimentation analysis
G. Electrochemical, or Nernst potential

Definition. The *Nernst potential* is determined by the *activity of each ion* in solution for which the surface of the particle is a *reversible* electrode. (The effects of the ions are additive). Therefore,

$$V = V_0 - (RT/nF) \cdot \ln a, \qquad (1\text{-}11)$$

where

V_0 = the Nernst potential when the activity of the ion a is 1.0 (*ideal* case)
R = gas constant (joules $\cdot$ mole^{-1} $\cdot$ deg^{-1})
T = temperature (K)
n = the number of electrons transferred in the reaction
F = 96,500 coulombs (Faraday's number).

H. Measuring the zeta potential

The *zeta potential* (that is, the potential drop across the movable part of the double layer), defined above at item C and in Figure 1-9, can be readily measured by one of the following methods:

    1. The gel electrophoretic method
       a) Electrophoretic mobility

Definition. The *electrophoretic mobility* of a particle is defined as the velocity of that particle relative to the dispersing medium, per unit of electrical potential gradient (volt/centimeter), or electric field **E** (Bier, 1967) (see Eq. (1-9)).

       b) Zeta potential and electrophoretic mobility

The relationship between the zeta potential and electrophoretic mobility can be represented by

$$\mu = \epsilon \zeta_0 \cdot f(\kappa r)/4 \cdot kT\eta \qquad (1\text{-}12)$$

where

μ = the mobility of the particle
η = viscosity of the medium
ϵ = dielectric constant
$f(\kappa r)$ = Debye function of $\kappa r = C \cdot \Sigma \, n_i Z_i^2$
r = radius of the particle extended to the plane of shear
κ = Debye–Hückel coefficient $\simeq$ "radius^{-1}" of the ion atmosphere
e^- = the charge of the electron

k = Boltzmann constant
T = the absolute temperature
n_i = number of ions of the ith species per cm^3
Z_i = the valence of the ion of the ith species, multiplied by e^-

2. Other electrophoretic methods
   a) Microscopic
   b) Free boundary (Whitney, 1977)
   c) Zonal

References

Becher, P. 1935. *Emulsions, Theory and Practice*, 2d ed. New York: Reinhold.

Bier, M. 1967. *Electrophoresis*, Vols. I and II. New York: Academic Press.

Einstein, A. 1911. New determination of molecular dimensions. *Ann. Phys.* **34**:591–592.

Flory, P. J. 1953. Configurational and frictional properties of the polymer molecule in dilute solutions. In *Principles of Polymer Chemistry*, 595–639. Ithaca, N.Y.: Cornell Univ. Press.

Morawetz, H. 1965. *Macromolecules in Solution*. New York: Wiley Interscience Publishers.

Mueller, H. 1943. The theory of electrophoretic migration. In *Proteins, Amino Acids and Peptides*, E. J. Cohen and J. T. Edsall, eds., 428–452. New York: Reinhold.

Scatchard, G., J. S. Coleman, and A. L. Shen. 1957. Binding of small ions to proteins. *J. Am. Chem. Soc.* **79**:12–16.

Scatchard, G., I. H. Scheinberg, and S. H. Armstrong. 1950. Physical chemistry of protein solutions. *J. Am. Chem. Soc.* **72**:535–540.

Tanford, C. 1961. *Physical Chemistry of Macromolecules*. New York: Wiley.

Whitney, R. McL. 1977. Chemistry of colloidal substances: General principles. In *Food Colloids*, H. D. Graham, ed., 14–21 and 56–59. Westport, Conn.: AVI.

2

Elements of Thermophysics and Chemical Thermodynamics: Basic Concepts

THERMOPHYSICS

Basic Concepts

In many practical situations, such as in a manufacturing process or in everyday life when one is running cars, trains, ships, or airplanes, a series of changes that are accompanied by *thermal* effects (heat exchange) occur in such machines. Such *thermal* changes are also very important in food processing. Macroscopic, or bulk, changes (*thermodynamic processes*) can be measured in such situations. If there are no changes in chemical structure, such as in the melting of ice, the processes are in the domain of physical thermodynamics or thermophysics. The study of both thermophysical and chemical processes is the subject of thermodynamics.

To determine the outcome of such a process, it is necessary to know the participants in the process, as well as their *mass* and *energy*. A certain region, or *system*, is therefore selected for study, and an inventory is made of its contents and the energy it exchanges with its surroundings, or the environment. When a process occurs, mass and/or energy may be transferred between the system and the environment at a *boundary* defined for the system under consideration. The major task of thermodynamics is to keep track of the *balance* of such exchanges and to relate them to changes in well-defined properties or variables of the system. Among the important variables considered are volume (V), pressure (P), mechanical work (W), and heat exchanged (Q_{exch}). These variables are commonly known from mechanics (V, P, W) or experimental work on heat transfer (Q_{exch}); they are not, however, sufficient to characterize a thermodynamic system. Thermodynamics, therefore, introduces several concepts that are essential to understanding both thermophysical and chemical processes.

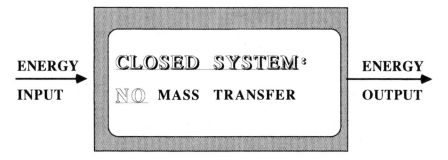

FIGURE 2-1.   Illustration of a thermodynamic, *closed* system, having (by definition) *only* energy transfer with the surroundings ("environment").

To simplify matters, classical thermodynamics, or "thermostatics," considers only systems that do *not* change their mass and have *no mass exchange* with the environment. Such systems are called *closed*, but they are still allowed to exchange energy with the environment (Fig. 2-1). An example of a closed system is a sealed tank containing a liquid or a gas.

Equilibrium States

If the mechanical forces between a closed system and the environment completely balance each other out, then the system is said to be in *mechanical equilibrium*. Furthermore, if the system does not exchange heat with its environment ($\delta Q_{exch} = 0.0$), and there are no processes inside the system involving heat exchanges, the system is said to be in *thermal equilibrium*. If, in addition, there is no net change in the chemical components of the system, then the system is in *chemical equilibrium* as well. Whenever the system is in chemical, thermal, *and* mechanical equilibrium, so that *no* chemical, thermal, or mechanical changes can occur *inside* the system, the system is said to be in *thermodynamic equilibrium*. In classical thermodynamics, or "thermostatics," the state of a system can only be defined *at equilibrium*.

If following the occurrence of a thermodynamic process it is possible to restore both the system and the surroundings to their initial states without spending energy, then the process is called *reversible*; in the opposite case, the process is *irreversible*. Thermostatics considers only reversible processes that are essentially a sequence, or succession, of equilibrium states; therefore, thermostatics is also called the thermodynamics of reversible processes or *equilibrium thermodynamics*. Most natural processes that occur frequently are, however, irreversible. In many cases, equilibrium thermodynamics is able to "substitute" a series of *very slow* changes that lead from the initial state to the final state of the process, in an almost reversible, or "quasi-static" manner, therefore allowing one to evaluate the overall energy exchanges in the closed system. When this can be done, the calculations of the thermodynamic quantities of interest are much simpler than for the original, *irreversible* process. Since in most cases one is concerned only with equilibrium, initial, and final states of a thermodynamic process, the simplicity of such calculations in the thermostatics approach makes it very attractive and useful. In the equilibrium state, the properties of the thermodynamic system are *constant* and spatially *uniform* throughout the system.

Temperature
We have seen that the variables of a thermodynamic system include volume (V), pressure (P), mechanical work (W), and heat exchanged (Q_{exch}), and that these variables are *not* sufficient to completely characterize the system. Practical experience leads one to the intuitive notions of "hotness" or "coldness" that need to be refined into a precisely formulated concept of *temperature*. In order to do so, let us consider a simple system, such as a single-component, single-phase liquid in equilibrium. To define the equilibrium state of the fluid, only two of its properties would have to be held constant. Let us assume that the fluid is liquid mercury held in a glass tube at constant pressure. If the liquid mercury is in equilibrium, only one other property could be varied while maintaining constant pressure. The volume (or height) of the mercury column in the glass tube observed at any equilibrium state will vary proportionally with the other system properties (except the pressure). The degree of "hotness" is then one such property that is measured against, or is proportional to, the *height* of the mercury column in the thermometer. This property is called *temperature*, and at this point, its scale, or range of values, is only defined empirically, and depends on the fluid employed in the thermometer tube to measure temperature.

Let us further consider a container with a partitioning wall, with liquid water on one side of the partition and alcohol on the other. Let us also assume that both the water and alcohol compartments have reached equilibrium. In this case, one finds experimentally that a thermometer will register exactly the same volume (height) when placed in either of the two compartments in equilibrium; this will be true for any thermometer. This experimental fact is formulated as the *zeroth law of thermodynamics*. If two systems A and B are separately in thermal equilibrium with a system C (the "thermometer"), then A and B are in equilibrium with each other. This law allows one practically to determine temperature (and equilibrium) with a thermometer, under conditions of constant pressure. Since the specification of an equilibrium thermodynamic state requires temperature to be determined, one says that temperature is a *state property*. For a given mass of a single-phase, single-component fluid system, the temperature is a function of pressure and volume:

$$T = T(P, V) \qquad (2\text{-}1)$$

Microscopically, the temperature is proportional to the degree of agitation or to the average rate of random, thermal motions of the molecules present in the system. The higher the temperature, the faster such motions are, *on average*, because the temperature is a bulk, or *macroscopic*/thermodynamic property.

The zeroth law, however, does *not* specify the scale of temperature, and if one were to measure temperature with thermometers containing *different* operating liquids, one would obtain different scales for temperature. Therefore, one would only have some *arbitrary* measure of the average degree of thermal agitation of the molecules in the system, which could lead to confusion.

If one were able to cool a fluid to the point where such thermal motions of molecules became absent or negligible, one would be then inclined to choose that point as the zero of the temperature scale, and one would be further inclined to consider that such a zero temperature point could be *the same* for all fluids or thermodynamic systems, and would be, therefore, *absolute*.

This proposition will be considered in more detail in connection with the following laws, or principles, of thermodynamics.

Three Thermodynamic Principles

The following three thermodynamic principles are the most important laws that govern the conversion of energy from one form to another, and the exchange of energy between the system and environment (in addition to the *zeroth law* of thermodynamics concerning the absolute zero of the temperature scale).

Internal Energy

Let us consider a special type of closed system that can neither gain energy nor lose it to its surroundings. Such a system is called *isolated*, and a definite amount of energy is trapped in the system, called *internal energy* (U). The internal energy consists of both the *kinetic* and *potential* energy of the trapped molecules. Like temperature, the internal energy of a system is a *variable of state*, and is *required* to specify the state of a system.

The First Principle: Conservation of Energy

First Principle. Energy is neither created nor destroyed; "the total energy of the world (system plus surroundings) is constant, or conserved."

A more detailed, precise formulation of the first law follows, relating the internal energy of a system to *both* mechanical work and heat exchange.

Consider a very large water bath (reservoir), in equilibrium with the surroundings, and that does *not* exchange heat or mechanical work. Within this large system, a tiny cylinder of gas is placed (Fig. 2-2), which is in thermal contact with the water and has a piston connected to the outside. This small system can exchange heat with the reservoir; also work, W_s, can be done on the small system by moving the piston. If there was no heat transfer between the small system and the reservoir, then the mechanical work, W_s, would be

$$W_s = U_s = -U_r$$

or

$$U_s + U_r = 0 \tag{2-2}$$

On the other hand, if the small system exchanged both heat (Q_s) and work (W_s) with the reservoir,

$$\Delta U = \delta Q_{\text{exch}} + W_{\text{exch}} \tag{2-3}$$

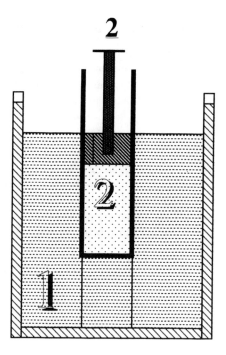

FIGURE 2-2.   Schematic illustration of a thermodynamic transfer process involving a large isothermal bath system (1) and a much smaller piston system (2). (Further details of the process and analysis are given in the text.)

that is, the algebraic sum of heat and work exchanged in a process is equal to the change in the internal energy state function, U. Equation (2-3) may also be considered as a more precise definition of the internal energy, U. This function does not depend on the path of the process (Fig. 2-3), but only on the state, and is therefore called a *state function*. The *path* of a process is the entire series of states through which the system passes when going from the *initial* state (i) to the *final* state (f).

Heat Capacity

Heat capacity is an especially important concept in thermochemical and heat-transfer calculations. The *heat capacity* is the amount of thermal energy that can be absorbed by a system for a rise in temperature by one degree (or unit):

$$\delta Q_{exch} = C_{process}\, dT \tag{2-4}$$

where $C_{process}$ is the heat capacity of the system for a certain type of process. This property of the system is *not* a state function since its value depends on the path, or on the process.

In a closed system, the internal energy change (written as an exact differential), dU, for a process occurring at constant pressure, is given by

$$dU = \delta Q - PdV = C_p dT - PdV \tag{2-5}$$

(Note that δQ is *not*, in general, an exact differential.)

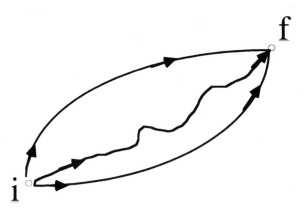

FIGURE 2-3.   Representation of *equivalent* pathways in a thermodynamic process beginning at the initial state (i) and ending in the final state (f).

For constant pressure processes it is therefore convenient to define a new state function, whose change in the process is

$$dH = dU + PdV = C_p dT \tag{2-6}$$

This new state function, $H = U + PV$, is called *enthalpy* (from the Greek words "en" and "thalpien," which mean "to warm" or "heat"). During an *isobaric* process (that is, at *constant pressure*), the enthalpy change is equal to the *heat transferred* during the process.

Another example is the *work* developed in an engine or turbine that runs steadily and does not exchange heat with the environment (called an *adiabatic process*, with $\delta Q_{exch} = 0.0$). This work is equivalent to the enthalpy change, ΔH, of the fluid passing through the engine.

In connection with equations (2-6) and (2-7), the heat capacity at constant pressure can be calculated as

$$C_p = \left(\frac{\partial H}{\partial T} \right)_p \tag{2-7}$$

where T is the temperature.

The Second Principle: Total Entropy Increases Always in a Closed System

All processes occurring in nature are in agreement with the first principle, that is, that total energy is conserved. However, there are many permitted processes, which although they would conform with this principle, as far as we can see, never occur. Therefore, there must be certain additional restrictions on the occurrence of processes that were not specified by the first principle. Consider, for example, a gas that initially fills just half the volume of a container, but is then allowed to diffuse very slowly into the whole container. Although the reverse process of the gas condensing itself into only half the volume would conserve total energy, this reversal does *not* occur naturally without causing some permanent change in the surroundings. Although energy is conserved when mechanical work is converted into heat, the conversion of *all* the heat back into mechanical work is *not* possible through *any reversible* process.

For a gas at low pressure and/or high temperature, it was experimentally found that the *universal gas law* holds:

$$PV = nRT \tag{2-8}$$

where n is the number of moles of gas, P is pressure, V is the volume occupied by the gas, and R is the universal gas constant. For a process at constant pressure one calculates the internal energy change of the gas from equation (2-5):

$$dU = \delta Q - PdV$$

or

$$dU = C_p dT - PdV$$

and replace PdV with $nRdT$ from equation (2-8). Therefore, $dU = \delta Q - nRdT$, and $dU = (C_p - nR) \cdot dT$. Alternatively,

$$\delta Q = dU + nRdT \qquad (2-9)$$

One has the interesting situation that the division by the temperature, T, on both sides of equation (2-9), yields a new state function that does *not* depend on the path

$$\frac{dU}{T} + nR\frac{dT}{T} = \frac{\delta Q}{T} = S(T) \qquad (2-10)$$

since both dT/T and dU/T can be integrated exactly. The new state function in equation (2-10) is called *entropy* and can also be calculated as

$$S(T) = \int_0^T C_p \cdot \frac{dT}{T} \qquad (2-11)$$

The entropy of a system (from the Greek words "en" and "tropos," which mean "to turn" or "change") when multiplied by temperature is a measure of the amount of *energy unavailable for work* during a natural process. Entropy increases continuously with the temperature and the degree of disorder, or randomness, in the system.

The *entropy change* for a reversible process is defined more generally as

$$\Delta S = \frac{\delta Q_{rev}}{T} \qquad (2-12)$$

With this definition, equation (2-5) can be rewritten as

$$dU = TdS - PdV = \delta Q + \delta W \qquad (2-13)$$

(Note that dS, unlike δQ, is always an *exact* differential.)

From equation (2-13) the heat exchanged can be calculated as

$$\delta Q = TdS - PdV - \delta W = TdS - \delta a \tag{2-14}$$

In a process in which heat transfer with the surroundings is zero (previously called *adiabatic*), $\delta a = TdS$ represents an internally generated thermal energy of the system. Because δa is *not* transferred from the surroundings, it is called *uncompensated heat* of the process. From equation (2-14) one also obtains

$$dS = \frac{\delta Q_{exch}}{T} + \frac{\delta a}{T} \tag{2-15}$$

where $\delta a / T$ is the entropy produced internally, dS_{int}, as a result of interconversion of mechanical work (eq. (2-15)), and $dS_{ext} = \delta Q_{exch}/T$ is the entropy change caused by the exchange of the heat δQ_{exch} with the surroundings. Equation (2-15) can also be written in the simpler form

$$dS = dS_{ext} + dS_{int} \tag{2-16}$$

In all spontaneous processes studied, it was found that the uncompensated heat, δa, is always *positive* or *zero*. This has been reformulated as the *second principle of thermodynamics*:

$$\boxed{dS_{int} \geqq 0} \tag{2-17}$$

as in equation (2-16),

$$dS = dS_{ext} + dS_{int}$$

where $dS_{int} > 0$ corresponds to *irreversible* processes, $dS_{int} = 0$ is valid *at equilibrium*, and $dS = \delta Q_{rev}/T$ holds for *reversible* processes, in general.

A simpler and more suggestive form of the second principle is usually stated for an isolated system (whose internal energy does not change, $\Delta U = 0.0$), for *any* process, spontaneous or otherwise. In this case, the surroundings do *not* contribute to the entropy of the isolated system, and $dS = dS_{int}$ (because $dS_{ext} = 0$). Because the uncompensated heat is always positive, for a spontaneous process the entropy increases up to a maximum value, at which point only reversible processes are possible.

In summary, the second principle states that entropy *cannot decrease* in a closed system.

As an example consider the arrangements of molecules shown in panels A and B of Figure 2-4 for a closed system. The regular arrangement in panel A

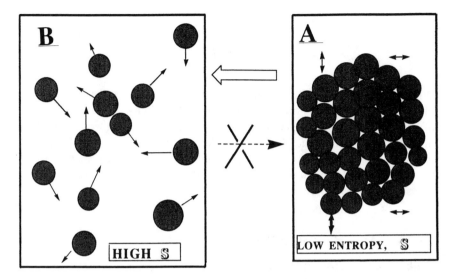

FIGURE 2-4.   Entropy-*forbidden*, "would-be" transition from a disordered state (A) to an ordered state (B) in a *closed* system. The reverse transition for which the entropy increases will spontaneously occur in a closed system.

of Figure 2-4 has lower entropy than the less regular arrangement in panel B of this figure, and therefore, the process B $\rightarrow$ A *cannot* occur in this *closed* system.

Third Principle: Entropy Tends to Zero at Zero Absolute Temperature

Both the first and second principle consider changes, or differences, in functions of state. The definition of entropy change (eq. (2-12)) links the entropy to temperature, but it does *not* define the scale for entropy. On the other hand, the zeroth law is concerned only with the zero point of the absolute temperature scale.

It was found experimentally that in many isothermal processes involving pure crystalline phases the entropy change tends to go toward zero as the temperature approaches zero degrees Kelvin (*absolute zero*, or 0 K).

Third Principle. The entropy of all pure crystalline phases at absolute zero temperature, $S(0 \text{ K})$, is equal to zero, $[S(0 \text{ K}) \rightarrow 0.0]$.

The third principle implies that the entropy can be determined from heat-capacity measurements at low temperatures (eq. (2-11)):

$$\Delta S = S(T, P) - S(0 \text{ K}) = S(T, P) \qquad (2\text{-}18)$$

or

$$\Delta S = \int_0^T (C_p/T) \, dT$$

Whereas the first two principles are virtually unconcerned about the *structure* of the system, the third principle considers the "perfectly" *ordered* structures of crystalline phases in the absence of thermal agitation (which would introduce disorder). The third principle also provides a useful link between thermodynamics, statistical mechanics, and spectroscopy (see Chap. 5), thus allowing the calculation of the thermodynamic properties from the spectroscopic properties of molecules (described in Chaps. 3 and 5).

CHEMICAL THERMODYNAMICS

Gibbs Free Energy, or Free Enthalpy

Many chemical reactions occur at constant temperature and pressure. For such reactions the enthalpy, $H = V + PV$, defined in equation (2-7) is convenient for calculations.

Considering that $\delta Q = TdS - \delta a$ (eq. (2-13)), $dU = TdS - \delta a + \delta W$ (eq. (2-12)), one can introduce a state function, the *free enthalpy*. Free enthalpy is defined as

$$G \equiv H - TS = U + PV - TS \tag{2-19}$$

which for a reversible process leads only to *mechanical work* terms:

$$dG = dU + PdV - TdS$$
$$= TdS - \delta a + \delta W + PdV - TdS \tag{2-20}$$

or

$$dG_{rev} = \delta W_{max} + PdV \tag{2-21}$$

since $\delta a = 0.0$. On the other hand, for a spontaneous process,

$$dG_{irrev} = -\delta a + \delta W + PdV \tag{2-22}$$

or

$$dG < 0 \tag{2-23}$$

whereas at equilibrium:

$$dG = 0.0 \tag{2-24}$$

and the *Gibbs free energy*, G, is minimum.

The quantity $\delta W_{net} = \delta W_{max} + PdV$ in equation (2-21) is the *net maximum work* that a system can perform in excess of the expansion work, under conditions of constant temperature and pressure in a *reversible* process.

Although it is closely related to enthalpy, G is often called *Gibbs free energy*, and it is very important for the study of chemical reactions, in general, as well as for biochemical reactions, or chemical reactions occurring in foods, in particular.

As a mnemonic device consider the thermodynamic "mountain" shown in Figure 2-5, which represents the three possibilities for a process: $\Delta G = 0$ (equilibrium), $\Delta G < 0$ (spontaneous process), or $\Delta G > 0$ (an "uphill" process that requires energy input to climb over the barrier).

Helmholtz Free Energy

For processes occurring at constant temperature, which are called *isothermal*, it is convenient to define a new state function, called *work function* or *Helmholtz function* (free energy):

$$A \equiv U - TS. \tag{2-25}$$

The Helmholtz free energy change at constant temperature is therefore defined as:

$$dA = -\delta a + \delta W - TdS \tag{2-26}$$

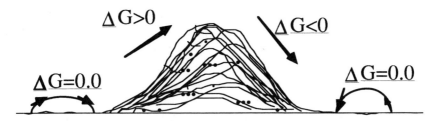

FIGURE 2-5.   Illustration of the Gibbs free-energy principle. For a system in thermodynamic equilibrium a *reversible* process *always* yields $\Delta G = 0$. Other processes are either *uphill* ($\Delta G > 0$; i.e., the system requires energy to go uphill) or *downhill* ($\Delta G < 0$ allows the process to occur spontaneously, or downhill) on the "thermodynamic mountain." Many different pathways, however, are possible for any ΔG value (see also Fig. 2-3 regarding pathways).

For a *spontaneous* process,

$$dA_{\text{irrev}} = -\delta a + \delta W \qquad (2\text{-}27)$$

On the other hand, for a *reversible* process,

$$dA_{\text{rev}} = \delta W_{\text{max}} \qquad (2\text{-}28)$$

which represents the *maximum mechanical work that the system can perform isothermally* (i.e., at constant temperature), since TdS is unavailable for work.

3

Elements of Quantum Mechanics and Quantum Chemistry

A *chemical bond* is caused by relatively strong attractive forces that hold atoms together in molecules and solids. The nature of the chemical bonds (Pauling, 1960), as well as the basis for understanding the mechanisms of chemical reactions and spectroscopic techniques, depends upon a theory of structure and processes at the electronic or submolecular level that is now called *quantum mechanics/quantum chemistry*. The concepts of chemical bonding, orbitals, electron distribution, and charge densities are all part of the treatment of molecular interactions and reaction mechanisms. These concepts are presented in some detail in this chapter.

The term *quantum* originates from the fact that the absorption or radiation of energy occurs in discrete packets called *quanta*. For example, a light quantum is a *photon*. Absorption or emission for photons results from the transitions between discrete electron energy levels in molecules. Therefore, quantum mechanics allows one to understand and quantitate the formation of atoms from nuclei and electrons. It also permits us to understand in detail the formation of molecules from atoms. Theoretical quantum calculations predict important properties of molecules, such as bond energies, bond lengths, electron distributions/dipole moments, and atomic or molecular spectra.

In the early stages of the development of the theory, the *Bohr model* (Bohr, 1913) of the atom was employed. It considered electrons moving around the positively charged nucleus in fixed, stable orbits. The properties of the bound electron, such as velocity, angular momentum, and mass at rest, were then estimated by employing Newtonian mechanics, combined with classical electromagnetism (Maxwell's equations). This oversimplified *fixed-orbit model* was later abandoned when bound electrons came to be considered to behave not only as particles, but as waves, *at the same* time. The principle proposed by *de Broglie* (1925) associated a wavelength (expressed in meters) to an electron "particle" of momentum, mv, according to the equation

$$\lambda = \hbar/mv, \qquad (3\text{-}1)$$

where v is the electron velocity (m/s), m is the rest mass of the electron (9.11 $\times$ 10^{-31} kg), and $\hbar$ is Planck's constant (6.626 $\times$ 10^{-34} J $\cdot$ s) divided by 2π. In 1927 Davisson and Germer verified experimentally this "wavelike" character of the electron by obtaining an *electron diffraction pattern* from a single crystal of nickel. As predicted by the wave theory, this diffraction pattern consisted of concentric rings (caused by the interference of the diffracted electron waves) around the transmitted electron beam. This data was also similar to the results one might have obtained with light waves (or photons) interfering after the waves had passed through a glass ball, as in the Newton's rings experiment.

Other subatomic "particles" such as the proton and neutron were also shown subsequently to exhibit wavelike character, and neutron diffraction became an important, powerful tool for structure determination. The de Broglie equation, (3-1), was also shown to apply to both protons and neutrons, even though these "particles," when at rest, are about 1000 times "heavier" than the electron. Because their rest masses are much greater than that of the electron, their corresponding associated "de Broglie wavelengths" are much shorter than those of the waves associated with the bound electrons. Equation (3-1) also shows that by increasing the velocity of an electron beam one can shorten the associated electron waves. This relation is now routinely used in electron microscopes to obtain high-resolution images, with wavelengths as short as 1 Å (10^{-8} cm or 10^{-10} m), corresponding to interatomic distances in crystals. Because of this *wave nature* of matter, which has been observed for *all* subatomic and nuclear particles, a formulation of quantum mechanics in terms of wave propagation and interference, called initially *wave mechanics*, was developed.

THE SCHRÖDINGER EQUATION

The Schrödinger formulation begins with a wavefunction, Ψ, that represents the amplitude of the electron wave. If the electron motion were a classical wave motion—which it is not—the wavefunction would simply represent the displacement of the electron from its "equilibrium position," much like the tops of water waves that are displaced from a lake surface. In the case of the electron, however, the function Ψ^2 gives the distribution of the electron in space, expressed as a *probability*, or chance of finding the electron at any point in space. The hydrogen atom, ^{1}H, is often used to illustrate the quantum mechanics calculations because of its simplicity, which allows one to obtain very accurate and complete analytical solutions of the *Schrödinger wave-equation*. The *bound* electron has a corresponding electronic wavefunction that depends on the x, y, and z coordinates of the electron, $\Psi = \Psi(w, y, z)$. It can be shown that the electron in the hydrogen atom can access only a certain number of *discrete* states characterized by definite energy values, E_n, and wavefunctions, Ψ_n, that are the only solutions of the Schrödinger wave-equation:

$$\hat{H}\Psi_n = E\Psi_n \qquad (3\text{-}2)$$

The *ground state* is defined as the state of lowest energy, E_0, and has a corresponding electron distribution. The remaining higher energy levels E_1, E_2, $\cdots$, E_n correspond to the *excited states* of the electron (Fig. 3-1), and the *electron density* or *distribution* is given by $[(\Psi_n(x, y, z)]^2$. This electron distribution determines the scattering of X-rays by the electrons in a molecule, and can therefore be derived from X-ray diffraction measurements on single crystals (see Chap. 5). The same electron distribution also determines the chemical reactivity of the molecule. $\hat{H}$ is a *functional operator* defined in the Appendix.

The energy of the electron in each electronic state in a molecule, such as H_2, CO_2, H_2O, can in principle be calculated from the appropriate wavefunction. The electronic energy, which can be calculated as a function of the internuclear distance, determines the stability of the molecule and the bond length (the result is qualitatively illustrated in Fig. 3-2). A stable molecule always corresponds to a minimum in the total energy (i.e., nuclear plus electron energy) that occurs for a certain internuclear distance (see Fig. 3-6), corresponding to the bond length.

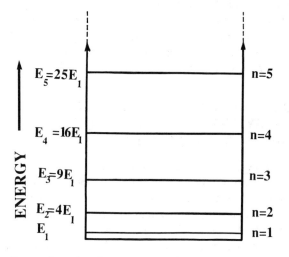

FIGURE 3-1.   Energy eigenvalues for a quantum particle in a one-dimensional potential well.

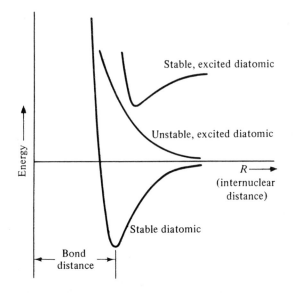

FIGURE 3-2.   Dependence of the potential energy of a molecule on the internuclear distance, R. The stable configuration at the bottom of the potential curve corresponds to the bond distance.

The calculation and measurements of energies of molecules are important in spectroscopy. A more detailed discussion of such measurements will be presented in Chapter 5.

For large molecules only approximate solutions for the Schrödinger wave-equation can be obtained, among which is the *Hückel molecular orbital theory*.

The quantum-mechanical operator, $\hat{H}$, in equation (3-6) plays a central role in all of quantum mechanics and quantum chemistry. Its form is analogous to the Hamilton operator (H) in classical mechanics, and is therefore routinely called a *Hamiltonian* in quantum mechanics (Mathews, 1968).

THE HYDROGEN ATOM AND THE SINGLE-ELECTRON SYSTEMS: He^+, Li^{2+}, Be^{3+}, AND OTHER SIMILAR SYSTEMS

The structure of the hydrogen atom (^1H) can be represented by the electrostatic attraction of an electron of negative charge, e^-, and mass, m_e ($|e^+| = |e^-| \equiv e$), and a proton of much larger mass, M_p (with $M_p \sim 1800M_e$). The electrostatic potential generated by this system of two opposite equal charges corresponds to a potential energy, $U = -e^2/r$, where r is the (variable) distance between the proton and the electron in the hydrogen atom. (In the SI system of units, $U = -e^2/(4\pi\epsilon \cdot r)$.) The electron distribution around the proton is specified by a wavefunction with space coordinates (x, y, and z, for example). Since the potential energy for this system of two charges depends only on the distance, r, between them, the corresponding wavefunction for the electron is *spherically symmetric*. In this case, the Schrödinger equation can be solved exactly (or analytically) to obtain both the wavefunctions and the corresponding energy levels for the single electron in the hydrogen atom. These energy levels are discrete, as illustrated in Figure 3-1, and can be written as E_0, E_1, $\cdot \cdot \cdot$, E_n. The index number, n, is called the *principal quantum number*, and reflects the dependence of the energy levels, E_n, on the distance between the proton and the electron. The wavefunction also has two other characteristic quantum numbers, the *angular momentum quantum number*, l, and the *magnetic quantum number*, m.

The permitted values of these three quantum numbers are as follows:

$$\left\{ \begin{array}{l} n = 1, 2, 3, \cdot \cdot \cdot \\ l = 0, 1, 2, \cdot \cdot \cdot, n - 1 \text{ (for a given value of } n) \\ m = -l, \cdot \cdot \cdot, 0, \cdot \cdot \cdot, \pm l \text{ (for a given value of } l) \end{array} \right.$$

The wavefunctions corresponding to $l = 0, 1, 2$, and 3 are labeled with the symbols s, p, d, and f, respectively. The fact that the energy levels of the hydrogen atoms are discrete, or distinct from each other, is reflected in the expression "the energy E of the hydrogen atom is *quantized*." The energy eigenvalues for the hydrogen atom (^{1}H) can be shown to be

$$E_n = \frac{-\mu e^4}{2h^2 \cdot n^2} \tag{3-3}$$

Although all three quantum numbers n, l, and m characterize the electron distribution (the "shape of the orbitals"), the energy eigenvalues only depend on the *principal* quantum number, n, as in equation (3-3), if the electron charge, e^-, is expressed in electrostatic units (esu), the mass μ is in grams, and h is in erg $\cdot$ s. On the other hand, if the energy value, E_n, is expressed in joules (J), for e^- in coulombs (C) and μ in kilograms (kg), then the *activity*, h, is in J $\cdot$ s.

Since the only variable in equation (3-3) is the quantum number, n, the remainder of the right-hand side of equation (3-3) can be replaced by a constant, R_b, called the *Rydberg constant:*

$$E_n = \frac{-R_b}{n^2} \tag{3-4}$$

where $R_b = 1312$ kJ/mol $= 2.179 \times 10^{-11}$ erg mol$^{-1} = 13.6$ eV/mol. The energy of the proton and electron considered separately, without the electrostatic interaction between them, is taken as the zero-reference level for the energy in the hydrogen atom. The value of R_b is therefore the *first ionization energy* of the H atom $(-13.6$ eV). The term R_b can be compared with other chemical bond energies and heats of chemical reactions that are routinely expressed in kJ/mol.

There is a whole series of single-electron systems (or species), such as He$^+$, Li^{2+}, Be^{3+}, that have very similar energy levels to those of the ^{1}H atom in equation (3-5). Since such systems have a charge Ze^+, with Z being the corresponding atomic numbers, the potential energy is $U = -Ze^2/r$ and the energy eigenvalues ("levels") are

$$E_n = \frac{-Z^2 R_b}{n^2} \tag{3-5}$$

The corresponding eigenfunctions, or wavefunctions, are hydrogenlike, and are therefore called *hydrogenic wavefunctions*. They are also characterized by the three quantum numbers, n, l, and m, as illustrated in Table 3-1.

THE ELECTRON DISTRIBUTION IN MOLECULES

Chemical properties of atoms or molecules are a direct reflection of the electron distribution in atoms or molecules. Furthermore, electric dipole moments of

TABLE 3-1   Hydrogenic Wave Functions*

n	l	m	
1	0	0	$\psi(1s) = \dfrac{1}{\sqrt{\pi}} \left(\dfrac{Z}{a_0}\right)^{3/2} e^{-Zr/a_0}$
2	0	0	$\psi(2s) = \dfrac{1}{4\sqrt{2\pi}} \left(\dfrac{Z}{a_0}\right)^{3/2} \left(2 - \dfrac{Zr}{a_0}\right) e^{-Zr/2a_0}$
2	1	0	$\psi(2p_z) = \dfrac{1}{4\sqrt{2\pi}} \left(\dfrac{Z}{a_0}\right)^{5/2} (z) e^{-Zr/2a_0}$
2	1	± 1	$\psi(2p_x) = \dfrac{1}{4\sqrt{2\pi}} \left(\dfrac{Z}{a_0}\right)^{5/2} (x) e^{-Zr/2a_0}$ $\psi(2p_y) = \dfrac{1}{4\sqrt{2\pi}} \left(\dfrac{Z}{z_0}\right)^{5/2} (y) e^{-Zr/2a_0}$

Source: Modified after Tinoco, Sauer, and Wang, 1978.

*$z = r \cos \theta$; $x = r \sin \theta \cos \phi$; $y = r \sin \theta \sin \phi$; $r^2 = x^2 + y^2 + z^2$; $a_0 = $ Bohr radius $= (h^2/me^2) = 0.529$ Å $= 5.29 \times 10^{-2}$ nm; $Z = $ charge on nucleus.

molecules, solvation of molecules, photochemical reactions, spectroscopic properties, and so on, all depend essentially on how the electrons are distributed in (or over) the molecule.

Somewhat disappointingly, very few (simplest) systems allow one to obtain exact solutions of the Schrödinger equation; as we have seen, the ^{1}H atom and the single-electron systems (He^+, Li^{2+}, Be^{3+}, $\cdots$) are in this category. However, relatively "simple" molecules, such as CO_2, H_2S, or H_2O, are still "too complicated" for the Schrödinger equation to give an exact calculation of their electron distribution. In such cases, as well as for larger molecules, models and approximation methods were developed based on quantum mechanics. Simplifications are often made by considering "localized" and "delocalized" electrons separately.

In all cases, the electron distribution determines the most stable configuration, or geometry, of a molecule, as well as other important molecular characteristics, such as bond lengths, bond angles, force constants, ionization potentials, and chemical reactivities.

For the hydrogen atom, the shapes of the wavefunctions (electron orbitals) can be computed as three-dimensional plots, as illustrated in Figure 3-3. The probability of finding the electron in a spherical shell of radius r/a_0 in the hydrogen atom can be calculated as $r^2\Psi^2$, where Ψ is the wavefunction and $a_0 = 0.529$ Å. Note the presence of distinct nodes in $r^2\Psi^2(2s)$ or $r^2\Psi^2(3s)$ (when the probability of finding the electron is zero), as illustrated in Figure 3-4.

The Hydrogen (H_2) Molecule and Many-Electron Atoms: Molecular Orbitals

The Hydrogen Molecule, H_2

Consider a hydrogen molecule (H_2), with the two protons (nuclei) labeled as A and B. The individual hydrogen atoms, in the absence of bonding between them, have the electron (atomic) orbitals illustrated in Figure 3-3. The electronic wavefunctions of the molecule (H_2) cannot be calculated exactly as a solution of equation (3-2), but they can be approximated by a weighted average, or *linear combination of the atomic orbitals* (LCAOs). Such *molecular orbitals* (MOs) can each accommodate only up to one electron pair.

The simplest LCAO of H orbitals gives an MO of the H_2 molecule:

$$\Psi_{H_2} = (1/\sqrt{2})[\Psi_A(1s) + \Psi_B(1s)] \tag{3-6}$$

which contains the $1s$ hydrogen atom orbital (AO) on the proton A and the $1s$ hydrogen AO on the proton B. The normalization constant is $(1/\sqrt{2})$, so that the maximum probability ($\sim \Psi^2$) remains 1.00. Another similar MO of H_2 can be generated by subtraction of the two atomic orbitals:

$$\Psi_{H_2}^1 = (1/\sqrt{2})[\Psi_A(1s) - \Psi_B(1s)] \tag{3-7}$$

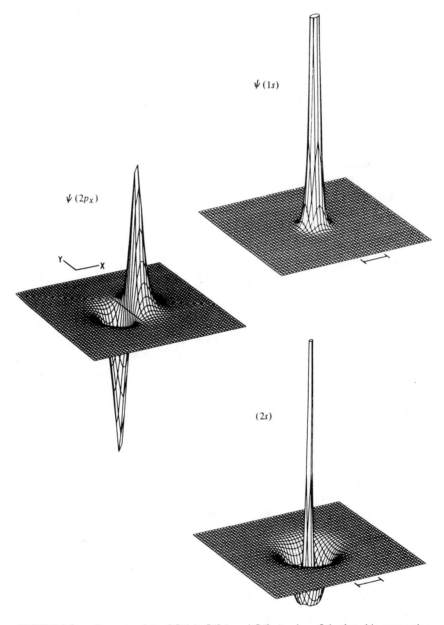

FIGURE 3-3.   Computer plots of $\Psi(1s)$, $\Psi(2s)$, and $\Psi(2p_x)$, where Ψ is plotted in perspective in the x-y plane. The marked line in front of the grid is of length $\ln a_0 = 5.29$ Å. (From A. Streiwieser, Jr., and P. H. Owens, 1973, with permission)

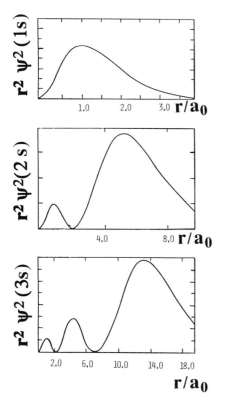

FIGURE 3-4.   Probability of finding the electron in a hydrogen atom in a spherical shell of radius r around the nucleus. The function $r^2\Psi^2$, which is proportional to this probabililty, is plotted vs. distance in units of $a_0 = 0.529$ Å (modified after Tinoco, Sauer, and Wang, 1978).

The approximate electron distributions corresponding to these MOs are obtained by calculating $\Psi^2_{H_2}$ and $(\Psi'_{H_2})^2$. These two MOs are labeled as σ and σ^*, respectively. The approximate energy levels of the electron for the H_2 molecule can then be computed with equation (3-1), but they are only relatively crude approximations of the "real" electronic energies of the H_2 molecule, since the electron interactions were neglected in the LCAO approximation. The σ-binding MO has cylindrical symmetry, with the H—H bond as the axis of symmetry and *no* node between the two hydrogens. The σ^*-antibonding orbital, on the other hand, has a node between the two hydrogens. There is a high probability of finding the electrons between the two protons, as shown by the electron density of the σ bond (Fig. 3-5a, b, and c). The result of this chemical bonding is the hydrogen molecule with a 0.78-Å bond length (1 Å $= 10^{-8}$ cm) and a *binding energy*, D, of 109 kcal per mole. Therefore, the hydrogen molecule is more stable than the two separate (or independent) hydrogen atoms by 109 kcal per mole. If one ignored the nuclear kinetic energy, then the *molecular (potential) energy* of H_2 has a variation upon the internuclear distance, as shown in Figure 3-6.

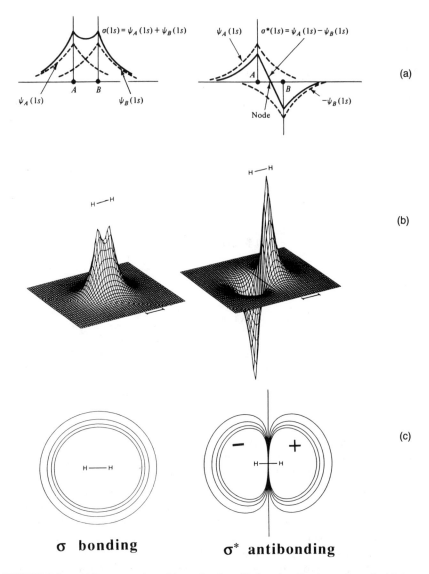

FIGURE 3-5.   (a) Representation of the molecular orbitals, σ-bonding (sum of two $1s$ AOs) and σ^*-antibonding (difference of two $1s$ AOs) for the H_2 molecule). (b) Computer-generated plots of the MOs. σ^* is the first excited state of the H_2 molecule. (Other conditions as in Fig. 3-3). (c) The $+$ and $-$ on σ^* MO indicate the signs of the corresponding wavefunctions. The marker is $2a_0 = 1.058$ Å for the grid in the computer plots (after Streiwieser and Owens, 1973).

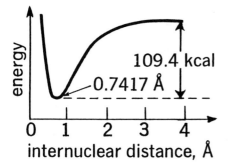

FIGURE 3-6. Quantitative illustration of the potential energy plot against internuclear distance for the hydrogen molecular gas according to the quantum chemical bond theory.

The function shown in Figure 3-7 can be determined experimentally, and is used to analyze/interpret the specific spectroscopic properties of the molecular hydrogen (H_2) gas. There are several other theoretical treatments of the H_2 molecule in addition to the MO theory, such as the *valence bond* (VB), the resonance, the self-consistent field (SCF) approximation (Kittel, 1963), and the hybridization method. The most accurate quantum chemical calculation to date for the chemical bond in H_2 involves a rather complicated wavefunction proposed by H. James, A. Coolidge, W. Kolos, and C. C. J. Roothaan (cited in Parker, 1983). This calculation agrees very well with all known properties of molecular (H_2) hydrogen; there is, in this case, an excellent agreement between the experimental (measured) value of the potential energy, D, and the calculated value.

Complex Molecules: π-Orbitals and Hybridization
For the next element after hydrogen, which is helium (He), there is no stable bond that is similar to that discussed previously for the H_2 molecule. Be_2 and Ne_2 are also unstable. The MO theory has been successfully applied to the more complex, diatomic molecules, such as N_2, O_2, F_2, and C=O. Molecular orbitals for the three homonuclear diatomic molecules are specified in Figure 3-7 *A* and *B*. In these molecules, a new type of MO, the π-orbital, is present; the π-orbitals are delocalized electron orbitals that do not possess the cylindrical symmetry of the σ-bonding orbitals (σ-bonding MOs). A π *(pi) -bonding molecular orbital* has a node along the line joining the two nuclei, and the electron density has a plane of symmetry passing through the bond, as schematically indicated in projection in Figure 3-8. In the case of the molecular (O_2) oxygen, optical excitation leads to the π*-orbital, and the double-bond character decreases upon excitation, which results in a weaker bond in the excited molecule.

To sum up simply the molecular orbital method, one considers a molecule as a *collection of molecular orbitals* of different symmetries (σ, σ*, π, π*, etc.). The distribution of electrons among these orbitals determines the type of chem-

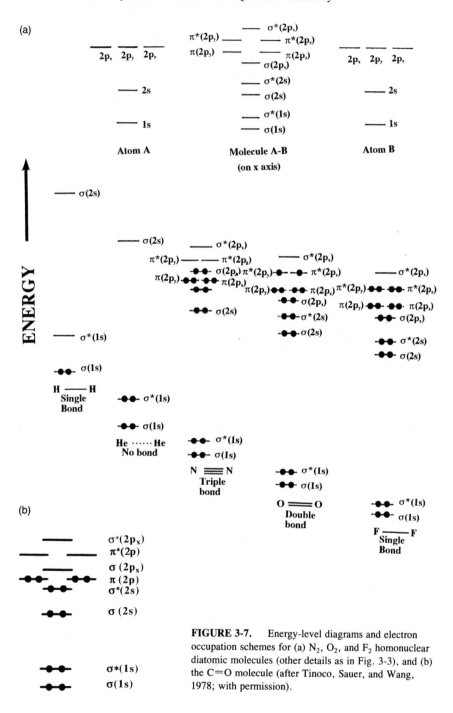

FIGURE 3-7. Energy-level diagrams and electron occupation schemes for (a) N_2, O_2, and F_2 homonuclear diatomic molecules (other details as in Fig. 3-3), and (b) the $C{=}O$ molecule (after Tinoco, Sauer, and Wang, 1978; with permission).

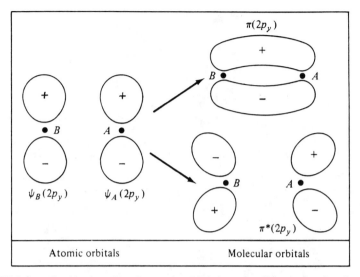

FIGURE 3-8.   Combination of two $2p$ atomic orbitals into a bonding π and antibonding π^* molecular orbital. Drawings represent the angular distributions of the electron densities; the $+$ and $-$ signs specify the sign of the wavefunction (after Tinoco, Sauer, and Wang, 1978).

ical bonding (single-, double-, or triple-bonding), as well as the stability of the molecule. Electrons are "placed" one at a time into the molecular orbitals that are spread over the entire molecule (Fig. 3-9). Such MOs are usually constructed as linear combinations of atomic orbitals (LCAOs). Finally, additional interactions are introduced between the MOs until the accurate potential energy curve (Fig. 3-6) is obtained.

To illustrate the level of accuracy currently attained by quantum–chemical computations, Table 3-2 presents the calculated and observed spectroscopic values for chemical bond properties of the more difficult CO molecule.

It must be emphasized, however, that the MO picture is, in general, an approximation (often a crude one, too!) that nevertheless helps systematize the experimental results. As indicated earlier, there are alternative approaches to treating chemical bonding. In the case of the O_2-molecule, for example, one can consider the chemical bonding on the basis of *hybridization* of atomic orbitals (AOs), before forming the MOs. This provides a refinement of the MO approximation.

Hybridization

The molecular orbitals of a multinuclear molecule can be, at least in principle, obtained from appropriate combinations of the atomic orbitals of the correct type of symmetry. The orbitals of a complex molecule are easier to visualize if

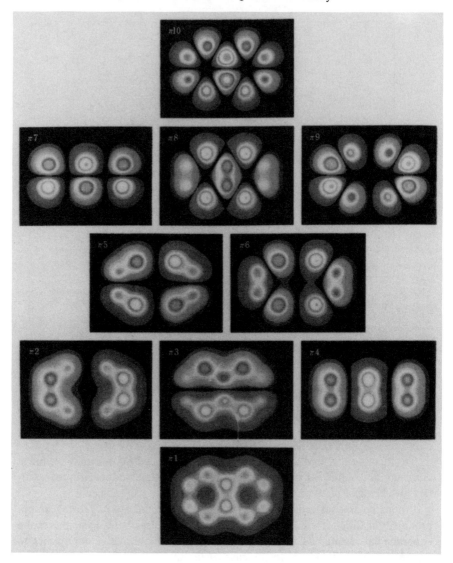

FIGURE 3-9.   π-Molecular orbitals of naphthalene.

the atomic orbitals are first combined into what is called a *set of hybrid orbitals*. These hybrid (or mixed) orbitals are then employed to form molecular orbitals with the neighboring atoms. The hybrid orbitals are suggested by the observed geometry of the molecule; for example, a 90° bond-angle is observed in H_2S without hybridization, whereas in H_2O, the bond angle calculated by employing

TABLE 3-2   Spectroscopic Properties of CO

Property*	Observed Value	Calculated Value†
r_e 10^8 cm	1.128	1.119
ω_e (cm^{-1})	2170	2357
$\omega_e x_e$ (cm^{-1})	13.5	11.1
B_e (cm^{-1})	1.93	1.97

*r_e = Internuclear distance at which the potential energy is minimum; ω_e = a quantity whose square is proportional to the curvature of the potential energy at its minimum; x_e = an harmonic constant; B_e = rotational constant.
†According to R. K. Nesbet (cited in S. P. Parker, ed., *McGraw-Hill Encyclopedia of Chemistry.* New York: McGraw-Hill, 1983, p. 136).

hybrid orbitals is 109.5° (the observed bond angle for H_2O is 104.5°). In CH_4, hybridization makes more orbitals available for bonding. There are numerous types of hybrid orbitals that could be considered, but for the sake of brevity, only three important types are considered here: the sp, sp^2, and sp^3 hybrids. The important difference among different hybrids is their *orientation* in three-dimensional space.

There are two sp hybrids formed from the possible combinations of a $2s$ and a $2p$ atomic orbital; there are three sp^2 hybrids formed by mixing a $2s$ with two $2p$ atomic orbitals; and there are four sp^3 hybrids made by mixing (or hybridizing) one $2s$ orbital with three $2p$ AOs. In the latter case all four sp^3 hybrids have identical shape, but they are oriented along different directions in three-dimensional space (they have different axes of symmetry). The angular dependence of an sp^3 hybrid orbital is illustrated in projection in Figure 3-10. Similar angular dependences are valid for sp and sp^2 hybrids, but their symmetry axes are differently oriented in three-dimensional space and their orientation closely approximates the observed molecular geometries. Examples of molecules whose

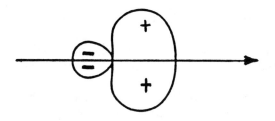

FIGURE 3-10.   Angular dependence of an sp^3 hybrid orbital; sp and sp^2 orbitals are similar. The arrow indicates the orientation of the hybrid. The plus and minus signs indicate the sign of the wavefunctions. A bond can be formed by combining this atomic orbital with an atomic orbital on another nulceus in the direction of the arrow (after Tinoco, Sauer, and Wang, 1978).

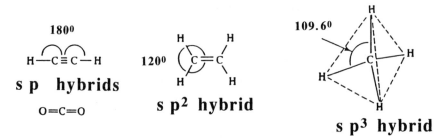

FIGURE 3-11.   Examples of molecules whose σ-bonds can be represented with hybrid atomic orbitals. In acetylene each C—H bond is a linear combination of an sp hybrid on C plus a $1s$ orbital on H. In CO_2 the sp hybrids on C interact with valence-shell orbitals on each O to form σ-bonds. In ethylene the six σ-bonds involve sp^2 hybrids on C. In methane the four σ-bonds are combinations of the four sp^3 hybrids on C, plus a $1s$ orbital on each H. The π-bonds in acethylene, CO_2, and ethylene are sums of $2p$ orbitals (modified after Tinoco, Sauer, and Wang, 1978).

σ-bonds can be represented by hybrid atomic orbitals are shown in Figure 3-11. In the case of the linear CO_2 (O=C=O) molecule, the two sp hybrids point in opposite directions, as demonstrated by Raman and IR spectroscopy. The bond angle between the two sp hybrids is 180°. On the other hand, a molecule such as ethylene ($H_2C=CH_2$), although it still involves a double bond, has three identically shaped sp^2 hybrids pointing toward the corners of an equilateral triangle; the bond angle between any two sp^2 hybrids is 120°, and all three orbital axes lie in the same plane, which means that the ethylene molecule is planar. Conversely, methane (CH_4) has four σ-bonds that can be represented as combinations of the four sp^3 hybrids centered on the carbon (C) atom, plus one $1s$ orbital on each hydrogen (H) atom. The water molecule, which is very important both in food and life sciences, also has sp^3 hybrids like methane; this fact is discussed in further detail in Chapter 4.

Mathematically, the wavefunction of an sp hybrid orbital is written as

$$\Psi(sp) = (1/\sqrt{2})\,\Psi_A(2s) + (1/\sqrt{2})\,\Psi_B(2p) \tag{3-8}$$

where $\Psi_A(2s)$ and $\Psi_B(2p)$ are the wavefunctions corresponding to the $2s$ and $2p$ orbitals, respectively. The orbital's angular dependence was shown in Figure 3-9.

In transition metals such as Fe, Cu, and Mo there are also d-orbitals present, and the d^2sp^3 hybrids result in octahedral, or *distorted octahedral*, coordination of six ligands. The electron distribution in such cases is localized between the participating atoms, and the approach taken is called the *Valence-Bond* (V-B) method.

In the case of organic molecules with delocalized electron distributions over several nuclei, the MO method discussed earlier is more appropriate. Real molecules are often more complex than either the V-B or MO pictures, but the V-B and MO calculations are helpful in providing details of molecular bonding and structure that could not be readily understood without them.

Conjugated Bonds and Delocalized Orbitals

In the previous section we considered MOs with each orbital localized around two neighboring atoms. This approximation is quite useful for σ-bonds, especially after hybridization for molecules such as CO_2, $H_2C=CH_2$, CH_4, and H_2O.

However, there are many organic molecules, such as the well-known (carcinogenic) *benzene* (C_6H_6) ring,

or the carotenoid pigments, which have *conjugated* π-bonds. In the classical (e.g., old organic) chemistry textbook notation, such bonds were represented as alternating double- and single-bonds: $\cdots -C=C-C=C-\cdots$, or $=C-C=C-C=$. It is well known that such representations of the bonds involved are inaccurate because the bond lengths are all the same (as in the benzene ring, according to Kékulé); also the corresponding bonding electrons are shared by the whole molecule because they are distributed, or *delocalized*, over the entire molecule. Such systems of conjugated π-electrons cannot be approximated by any hybrids centered on individual nuclei. In this sense, as well as in chemical reactivity terms, the sp^2 hybrid molecules of ethylene and benzene are quite different. In the conjugated system, the π-electrons are free to move throughout. Thus they can be treated as a "particle-in-a-box" model, whereas the entire molecule is considered as a box, or a trap, within which the π-electrons move freely. In the case of the benzene ring (Fig. 3-12), the π-delocalized electron orbitals form two rings (in the form of a doughnut, or *torus*) below and above the plane of the carbon atoms in benzene. These two π-orbitals force the benzene molecule into the planar configuration or geometry, and prevent the carbon atoms from moving out of the plane, except for very small out-of-plane vibrations. The partial double-bond character of the $C\underline{\cdots}C$ bonds in conjugated systems can be thought of as being the result of π-electrons occupying the two lowest energy orbitals, π_1 and π_2, with the central pair $C\underline{\cdots}C$ having less double-bond character than either of the two end $C-C$ bonds in a linear conjugated system. This is consistent with the experimental bond lengths of 1.476-Å length for the middle $C\underline{\cdots}C-$ bond and the 1.337 Å for the end $C\underline{\cdots}C$ bonds. The $\widehat{CCC}$ bond angle in a conjugated system is 122.9°, which is close to that required for sp^2 hybrids to form the σ-bond.

In terms of wavefunctions, the butadiene molecule, "$H_2C_a=C_bH-$

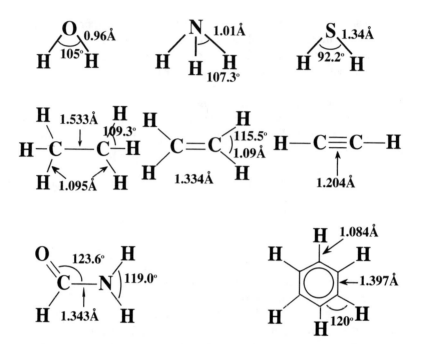

FIGURE 3-12.   Molecular geometry of several organic molecules. Bond angles are given in degrees, and bond distances are in angstroms (1 Å = 0.1 nm).

$C_cH=C_dH_2$'', can be treated by assigning each carbon an atomic $2P_z$ orbital, labeled as Ψ_a, Ψ_b, Ψ_c, and Ψ_d, respectively. The four (normalized) π molecular orbitals of butadiene can then be shown to be written as

$$\begin{cases} \Psi(\pi_1) = 0.372\Psi_a + 0.602\Psi_b + 0.602\Psi_c + 0.372\Psi_d \\ \Psi(\pi_2) = 0.602\Psi_a + 0.372\Psi_b - 0.372\Psi_c - 0.602\Psi_d \\ \Psi(\pi_3) = 0.602\Psi_a + 0.372\Psi_b - 0.372\Psi_c + 0.602\Psi_d \\ \Psi(\pi_4) = 0.37\Psi_a - 0.602\Psi_b + 0.602\Psi_c - 0.372\Psi_d \end{cases} \tag{3-9}$$

This is obviously an LCAO method, sometimes known as the HMO (Hückel molecular orbital). The HMO orbitals and the results for benzene, formamide, and the peptide bond (which all involve π-delocalized MOs) are given in Figures 3-13, 3-14, 3-15, respectively. (These were calculated on an Apple-Macintosh II [68020/68882, CPU/Math Coprocessor] with a program from Drexel University, readily available through Kinko's Company as Academic shareware.)

Secular Matrix

	1	2	3	4	5	6
1	0	1.0	0	0	0	1.0
2	1.0	0	1.0	0	0	0
3	0	1.0	0	1.0	0	0
4	0	0	1.0	0	1.0	0
5	0	0	0	1.0	0	1.0
6	1.0	0	0	0	1.0	0

BENZENE

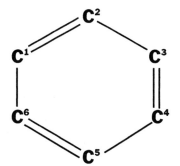

Number of π electrons: 6

Energy Eigenvalues

$E_1 = \alpha + 2.0000\beta$
$E_2 = \alpha + 1.0000\beta$
$E_3 = \alpha + 1.0000\beta$
$E_4 = \alpha - 1.0000\beta$
$E_5 = \alpha - 1.0000\beta$
$E_6 = \alpha - 2.0000\beta$

Eigenvectors

$\Psi_1 = 0.4082\phi_1 + 0.4082\phi_2 + 0.4082\phi_3 + 0.4082\phi_4 + 0.4082\phi_5 + 0.4082\phi_6$

$\Psi_2 = 0.5774\phi_1 + 0.2887\phi_2 - 0.2887\phi_3 - 0.5774\phi_4 - 0.2887\phi_5 + 0.2887\phi_6$

$\Psi_3 = 0.0000\phi_1 + 0.5000\phi_2 + 0.5000\phi_3 + 0.0000\phi_4 - 0.5000\phi_5 - 0.5000\phi_6$

$\Psi_4 = 0.0000\phi_1 - 0.5000\phi_2 + 0.5000\phi_3 - 0.0000\phi_4 - 0.5000\phi_5 + 0.5000\phi_6$

$\Psi_5 = 0.5774\phi_1 - 0.2887\phi_2 - 0.2887\phi_3 + 0.5774\phi_4 - 0.2887\phi_5 + 0.2887\phi_6$

$\Psi_6 = 0.4082\phi_1 - 0.4082\phi_2 + 0.4082\phi_3 - 0.4082\phi_4 - 0.4082\phi_5 - 0.4082\phi_6$

Charge Parameters

Atom	π Charge Density	Charge
1	1.0000	0.0000
2	1.0000	0.0000
3	1.0000	0.0000
4	1.0000	0.0000
5	1.0000	0.0000
6	1.0000	0.0000

Bond Parameters

Atom Pair	π Bond Order
1-2	0.6667
1-6	0.6667
2-3	0.6667
3-4	0.6667
4-5	0.6667
5-6	0.6667

FIGURE 3-13.   Hückel molecular orbital calculations for benzene.

Secular Matrix

	1	2	3
1	0	0.8	1.0
2	0.8	1.5	0
3	1.0	0	1.0

FORMAMIDE

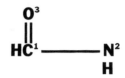

Number of π electrons: 4

Energy Eigenvalues

$E_1 = \alpha + 2.0667\beta$
$E_2 = \alpha + 1.2570\beta$
$E_3 = \alpha - 0.8237\beta$

Eigenvectors

$\Psi_1 = 0.5082\phi_1 + 0.7174\phi_2 + 0.4764\phi_3$
$\Psi_2 = 0.1925\phi_1 - 0.6339\phi_2 + 0.7491\phi_3$
$\Psi_3 = 0.8394\phi_1 - 0.2890\phi_2 - 0.4603\phi_3$

Charge Parameters

Atom	π Charge Density	Charge
1	0.5907	0.4093
2	1.8330	0.1670
3	1.5763	-0.5763

Bond Parameters

Atom Pair	π Bond Order
1-2	0.4852
1-3	0.7727

FIGURE 3-14.   Hückel molecular orbital calculations for formamide.

The structure of formamide is particularly informative in relation to that of the peptide bond. The H—N—H angle of 119° implies trigonal bonding around the nitrogen, and *not* tetrahedral bonding as in NH_3. The O—C—N angle is also trigonal, as expected. The C⸺N bond length is 1.343 Å, which is shorter than the C—N single bond length of 1.470 Å in methyl amine. The C⸺N bond therefore has partial double-bond character, and the structure of formamide is found to be planar (Fig. 3-12), which is in agreement with the partial double-bond character. In the MO picture, the π-orbitals of formamide would be delocalized over the three nuclei, C, N, and O and would force the C, N, and O into the same plane, with *no* rotation being permitted around the

Secular Matrix

	1	2	3	4	5
1	0	0.8	1.0	1.0	0
2	0.8	1.5	0	0	0.8
3	1.0	0	1.0	0	0
4	1.0	0	0	0	0
5	0	0.8	0	0	0

Number of pi electrons: 6

Energy Eigenvalues

$E_1 = \alpha + 2.3309\beta$
$E_2 = \alpha + 1.4662\beta$
$E_3 = \alpha + 0.4191\beta$
$E_4 = \alpha - 0.3200\beta$
$E_5 = \alpha - 1.3961\beta$

Eigenvectors

$\Psi_1 = 0.4963\phi_1 + 0.7137\phi_2 + 0.3729\phi_3 + 0.2129\phi_4 + 0.2449\phi_5$
$\Psi_2 = 0.3191\phi_1 - 0.5428\phi_2 + 0.6845\phi_3 + 0.2176\phi_4 - 0.2962\phi_5$
$\Psi_3 = 0.3148\phi_1 - 0.0966\phi_2 - 0.5419\phi_3 + 0.7510\phi_4 - 0.1843\phi_5$
$\Psi_4 = 0.0803\phi_1 + 0.3576\phi_2 - 0.0608\phi_3 - 0.2509\phi_4 - 0.8939\phi_5$
$\Psi_5 = 0.7392\phi_1 - 0.2426\phi_2 - 0.3085\phi_3 - 0.5294\phi_4 + 0.1390\phi_5$

PEPTIDE BOND

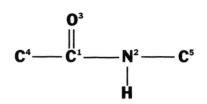

Charge Parameters

Atom	π Charge Density	Charge
1	0.8944	0.1056
2	1.6266	0.3734
3	1.8023	-0.8023
4	1.3134	-0.3134
5	0.3634	0.6366

Bond Parameters

Atom Pair	π Bond Order
1-2	0.3012
1-3	0.4658
1-4	0.8230
2-5	0.7067

FIGURE 3-15.   Hückel molecular orbital calculations for the peptide bond.

partial double bonds C⸱⸱⸱N and C⸱⸱⸱O. The peptide bond, too often written as

or

$$O=\overset{|}{\underset{|}{C}}-NH$$

is very similar to formamide. Its structure is planar because of the π-delocalized MOs over the C, N, and O nuclei, with both C⸺N and C⸺O having partial double-bond character. Again, the classical formula

fails to represent chemical reality and should, therefore, be written as

The consequences of this rigidity of the peptide bond for protein conformations will be discussed in detail in Chapter 4.

General Approach to Many-Electron Atoms and Molecules

For systems that contain more than one electron the Schrödinger equation has the form

$$\left[-\frac{\hbar^2}{2\mu} \sum_i \Delta_i - 2 \sum_i (e^2/r_i) - \sum_{ij} (e^2/r_{ij}) \right] \Psi = E\Psi \qquad (3\text{-}10)$$

where Δ_i are the Laplace (second-order differential) operators for the individual electrons i, μ is the reduced mass of the electron, and the indices i and j label all the different electrons in the system, with r_{ij} being the distance between electrons i and j. Because of the electron interaction term that contains r_{ij}, it is *not* possible to simplify Schrödinger's equation like we can in the case of the hydrogen atom (see the first section in this chapter). Therefore, an exact analytical solution of equation (3-10) is generally impossible. This means that one cannot write exactly the precise functional form of the wavefunction Ψ for atoms or molecules containing more than one electron (called here *many-electron systems*). The first such case is the He atom. However, the energy levels and electron distributions can still be determined by approximation methods with a high degree of accuracy. The calculated values agree in most cases with the measured values within the experimental error and the limits imposed by the *Heisenberg uncertainty principle*. Some of these methods (such as the MO and V-B methods) were discussed in the previous two sections, as well as by Coulson (1982) and Schaefer (1972).

The many-electron atoms are considered to have hydrogenlike wavefunctions and energy levels. The energy levels of many-electron atoms are, however, quite different from those of hydrogen atom. The *Pauli exclusion principle* applies, that is, *there are at most two electrons in each orbital, but no more.* Furthermore, the two electrons in a filled orbital must have opposite spins and are paired ($\uparrow\downarrow$). Pauli found that *no* two electrons in an atom can have the same four quantum numbers, n, l, m, and S. Two electrons in the same orbital have the same n, l, and m quantum numbers; therefore, they must have different spin numbers, S. Since S has only two values, $-1/2$ and $+1/2$, there are only two electrons that can have these spin numbers, and their spins are therefore antiparallel ($\uparrow\downarrow$). Following this principle, the electronic structures of all atoms in the periodic table can be specified. Two examples, for He and C, are shown in Figure 3-16.

Charge-Distribution and Dipole-Moment Calculations

An important use of the approximate wave functions is to estimate the charge distribution and dipole moments of molecules. This type of calculation can have as far-reaching consequences as the improved understanding of chemical carcinogenesis (Pullman and Pullman, 1969) and the design of new pharmaceutical compounds.

The charge distribution is calculated with the wavefunction by integration of the probability of finding the electron in a finite volume, V, of space:

$$\rho_{ch} = \int_0^V [\Psi_n(x, y, z)]^2 \, dx \cdot dy \cdot dz \tag{3-11}$$

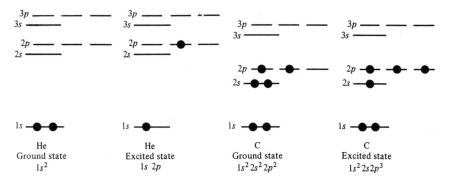

FIGURE 3-16.   Energy ordering and electron orbital notation for many-electron atoms. At most, two electrons are placed in each orbital in accord with the Pauli exclusion principle. The electronic state of the atom is specified by the number of electrons in each orbital. For example, the fluorine ground state is $1s^2 2s^2 2p^5$ (modified after Tinoco, Sauer, and Wang, 1978).

The HMO approach discussed in the previous section is often employed for this purpose. Results of such calculations for tryptophan and adenine are shown in Figure 3-17. The charge densities obtained are consistent with simple chemical reasoning, but there are two major sources of error in such approximate calculations: (1) the effect(s) of the solvent, and (2) the fact that the wavefunctions employed at the start of the calculations are only crude approximations of the "real" MOs. In the case of adenine, the calculation predicts that the three unsubstituted ring nitrogens are more negative than the amino nitrogen; this result therefore predicts that protonation occurs first on the ring. This prediction is verified experimentally. Since the charge densities can be measured, the propriety of the wavefunctions being employed can be checked.

The dipole-moment calculation is related to the charge distribution. The *dipole moment*, μ, determines the polarity of a molecule; thus, if the center of positive charge of a molecule is separated by a distance r_0 from the center of

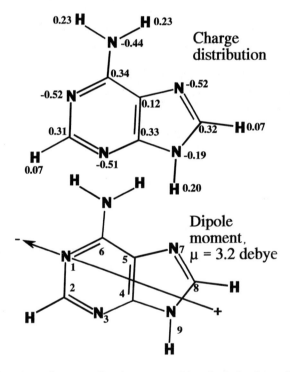

FIGURE 3-17.   Approximate wavefunctions were used to estimate the charge distribution and dipole moment of adenine. One debye is 10^{-18} esu $\cdot$ cm. The charge distribution is given as the fraction of an electronic charge at each atom relative to the charge (equal to zero) on the isolated atom.

negative charge from all the electrons in the molecule, then the molecule is polar and has a nonzero dipole moment. The dipole moment is, by definition,

$$\mu = q \cdot \mathbf{r}_0 \tag{3-12}$$

where $+q$ is the positive charge separated by the distance $\mathbf{r}_0$ from the negative charge, $-q$. The dipole moment is a *vector quantity* since it has both magnitude and direction (defined from the center of the positive charge to the center of the negative charge). Its unit is the debye (D) or 10^{-18} esu $\cdot$ cm. If the centers of positive and negative charge coincide, the value of $\mathbf{r}_0$ is zero and the dipole moment is also zero (as seen from eq. (3-12)). It is for this reason that molecules with point, or center, symmetry, such as H_2, N_2, O_2, CH_4, and benzene, all have *zero* dipole moments. These molecules are therefore *nonpolar*, or *hydrophobic*. A molecule with a plane of symmetry, such as the water molecule, tryptophan, or adenine, has a dipole moment located in the plane of the molecule. The magnitude of the dipole moment, $|\mu|$, can be determined experimentally from dielectric measurements on solutions containing the molecule. It can also be calculated from the charge-density distribution of the molecule by taking into account the positions of the nuclei and the charge density at each nucleus in the molecule, and then by applying equation (3-12) for the centers of positive and negative charge. The calculated dipole moments for adenine and water are also shown in Figure 3-13. Since the electron charge is 4.8×10^{-10} esu, and the internuclear distances are of the order of 10^{-8} cm (1 Å), the dipole moments are of the order of 10^{-18} esu $\cdot$ cm (1 debye (D) = 10^{-18} esu $\cdot$ cm, or 1D = 3.33564×10^{-30} C $\cdot$ m, in the SI system of units, with the electronic charge expressed in coulombs (C)). The adenine has a calculated dipole moment, $\mu = 3.2$ D, oriented as shown in Figure 3-17.

Interactions between large molecules such as proteins (which may have large dipole moments) can have quite significant effects on their solution properties. Nonideal behavior may occur as a result of such interactions. Dipole–dipole interactions will be discussed in some detail in Chapter 6.

Quantum Chemistry, Molecular Structure, and Molecular Orbitals: Brief Summary

Although the quantum chemistry, MO approach is only approximate, it is quite helpful for gaining an intuitive picture of chemical structure and reactivity. There are two basic assumptions in this approach that are both important and supported by (approximate) agreement with experimental data.

1. All bonding molecular orbitals can be expressed as a sum of hydrogenlike atomic orbitals on the different nuclei. (Nonbonding MOs are localized on a single nucleus.)

2. The electron distribution of any molecule can be calculated with a set of molecular orbitals and their corresponding energy levels. Hybrid orbitals are often more appropriate for calculating molecular structure (bond angles and bond lengths). The symmetry of hybrid orbitals is reflected in the symmetry of the molecule. Conjugated systems (peptide bonds, carotenoids/pigments, formamide, benzene, certain cancer-causing chemicals ("carcinogens")) possess π-delocalized orbitals that cannot be considered as belonging to only one or two nuclei; in several cases (e.g., benzene, carotenoids) such π-electrons are delocalized over the entire molecule, that is, they are shared by the entire conjugated molecule.

The conjugated-bond character is partial double-bond. The consequences of π-orbital delocalization are very significant both in terms of molecular geometry (stereochemistry) and chemical reactivity. Hückel molecular orbital results were employed to illustrate the method.

Approximate charge-density distributions and dipole moments can be determined with the MO approach, and are in reasonable agreement with experimental values for solutions. Hückel molecular orbital calculations are illustrated for adenine in Figure 3-18.

Pauli's exclusion principle leads to a representation of the electron configuration of all elements in terms of a set of electron orbitals with at most two electrons each.

Quantum mechanics/quantum chemistry serve as the essential theory for powerful, spectroscopic techniques; spectroscopic techniques and their applications will be presented in Chapter 5 and in Volume II of this book.

APPENDIX

The Hamiltonian: The Correspondence, Uncertainty, and Complementarity Principles

For very large numbers of atoms or molecules, quantum mechanics must take the simpler form of classical mechanics. This idea can be stated rigorously as follows:

> **The Correspondence Principle.** The relations between the essential physical variables (such as positions and momenta) in classical mechanics should have the same form for the quantum operators corresponding to these variables. (For a brief treatment of functional operators see also chap. 2, pp. 12–15, of Mathews, 1968).

In quantum mechanics the position operators are denoted as $\hat{x}$, $\hat{y}$, and $\hat{z}$, whereas the components of the *momentum operator* are denoted by $\hat{p}_x$, $\hat{p}_y$, and

Secular Matrix

	1	2	3	4	5	6	7	8	9	10
1	0.5	1.0	0	0	0	1.0	0	0	0	0
2	1.0	0	1.0	0	0	0	0	0	0	0
3	0	1.0	0.5	1.0	0	0	0	0	0	0
4	0	0	1.0	0	1.0	0	0	0	0.8	0
5	0	0	0	1.0	0	1.0	1.0	0	0	0
6	1.0	0	0	0	1.0	0	0	0	0	0.8
7	0	0	0	0	1.0	0	0.5	1.0	0	0
8	0	0	0	0	0	0	1.0	0	0.8	0
9	0	0	0	0.8	0	0	0	0.8	1.5	0
10	0	0	0	0	0	0.8	0	0	0	1.5

Number of pi electrons: 12

Energy Eigenvalues

$E_1 = \alpha + 2.6181\beta$

$E_2 = \alpha + 2.1148\beta$

$E_3 = \alpha + 1.7250\beta$

$E_4 = \alpha + 1.4083\beta$

$E_5 = \alpha + 1.0411\beta$

$E_6 = \alpha + 0.6397\beta$

$E_7 = \alpha - 0.7717\beta$

$E_8 = \alpha - 0.9807\beta$

$E_9 = \alpha - 1.2435\beta$

$E_{10} = \alpha - 2.0512\beta$

Charge Parameters

Atom	π Charge Density	Charge
1	1.2759	-0.2759
2	0.8621	0.1379
3	1.2704	-0.2704
4	0.9123	0.0877
5	1.0558	-0.0558
6	0.8433	0.1567
7	1.3019	-0.3019
8	0.8494	0.1506
9	1.7367	0.2633
10	1.8921	0.1079

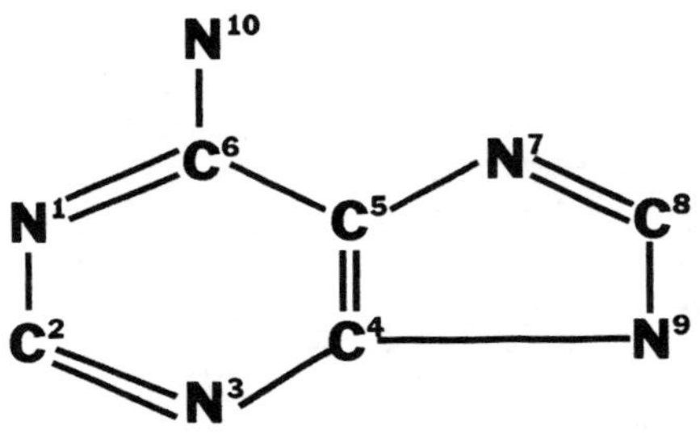

Bond Parameters

Atom Pair	π Bond Order
1 - 2	0.6338
1 - 6	0.6286
2 - 3	0.6773
3 - 4	0.5783
4 - 5	0.5895
4 - 9	0.3905
5 - 6	0.5601
5 - 7	0.4562
6 - 10	0.3475
7 - 8	0.7895
8 - 9	0.4498

ADENINE

FIGURE 3-18a.   Hückel molecular orbital calculations for adenine: energy eigenvalues, secular matrix, π-charge density, and bond parameters.

Eigenvectors

$$\Psi_1 = 0.2365\phi_1 + 0.1977\phi_2 + 0.2812\phi_3 + 0.3978\phi_4 + 0.3838\phi_5$$
$$\quad + 0.3033\phi_6 + 0.3039\phi_7 + 0.2599\phi_8 + 0.4705\phi_9 + 0.2170\phi_{10}$$

$$\Psi_2 = 0.3706\phi_1 + 0.1852\phi_2 + 0.0211\phi_3 - 0.1512\phi_4 + 0.0732\phi_5$$
$$\quad + 0.4132\phi_6 - 0.1073\phi_7 - 0.2465\phi_8 - 0.5174\phi_9 + 0.5377\phi_{10}$$

$$\Psi_3 = 0.2780\phi_1 + 0.4704\phi_2 + 0.5333\phi_3 + 0.1829\phi_4 - 0.1327\phi_5$$
$$\quad - 0.1228\phi_6 - 0.2821\phi_7 - 0.2128\phi_8 - 0.1063\phi_9 - 0.4615\phi_{10}$$

$$\Psi_4 = 0.0523\phi_1 - 0.0070\phi_2 - 0.0622\phi_3 - 0.0495\phi_4 + 0.4051\phi_5$$
$$\quad + 0.0545\phi_6 + 0.5654\phi_7 + 0.1085\phi_8 + 0.5157\phi_9 - 0.4758\phi_{10}$$

$$\Psi_5 = 0.5965\phi_1 + 0.1708\phi_2 + 0.4187\phi_3 + 0.3974\phi_4 - 0.2262\phi_5$$
$$\quad + 0.1520\phi_6 + 0.0098\phi_7 + 0.2315\phi_8 + 0.2891\phi_9 - 0.2650\phi_{10}$$

$$\Psi_6 = 0.0940\phi_1 + 0.3275\phi_2 + 0.3035\phi_3 + 0.2851\phi_4 + 0.3773\phi_5$$
$$\quad + 0.3406\phi_6 - 0.3843\phi_7 - 0.4310\phi_8 - 0.1357\phi_9 - 0.3167\phi_{10}$$

$$\Psi_7 = 0.0423\phi_1 + 0.4171\phi_2 + 0.3642\phi_3 + 0.0461\phi_4 + 0.1884\phi_5$$
$$\quad - 0.4710\phi_6 + 0.2795\phi_7 - 0.5438\phi_8 - 0.1753\phi_9 + 0.1659\phi_{10}$$

$$\Psi_8 = 0.4874\phi_1 - 0.3322\phi_2 - 0.1616\phi_3 + 0.5714\phi_4 - 0.2059\phi_5$$
$$\quad + 0.3894\phi_6 + 0.0199\phi_7 + 0.1764\phi_8 - 0.2412\phi_9 + 0.1256\phi_{10}$$

$$\Psi_9 = 0.2108\phi_1 - 0.4597\phi_2 - 0.3608\phi_3 - 0.1694\phi_4 + 0.3039\phi_5$$
$$\quad - 0.0922\phi_6 - 0.4551\phi_7 - 0.4896\phi_8 + 0.1922\phi_9 - 0.0269\phi_{10}$$

$$\Psi_{10} = 0.2798\phi_1 - 0.2707\phi_2 + 0.2754\phi_3 - 0.4319\phi_4 + 0.5494\phi_5$$
$$\quad - 0.4432\phi_6 + 0.2518\phi_7 + 0.0930\phi_8 + 0.0763\phi_9 + 0.0998\phi_{10}$$

FIGURE 3-18b. Hückel molecular orbital calculations for adenine: eigenvectors.

$\hat{p}_z$, along the corresponding x, y, and z directions. For a particle of mass m and momentum p, moving in a potential $V(x)$, one defines a function, called the *Hamiltonian*

$$\mathcal{H} = \frac{p^2}{2m} + V(x) \qquad \text{(A-1)}$$

that is conserved during motion and gives the total energy of the system. The expression $(p^2/2m) = mv^2/2$ represents the *kinetic energy* of the particle, since, by definition, the momentum is $\mathbf{p} = m\mathbf{v}$, where $\mathbf{v}$ is the velocity of the particle.

The Hamiltonian *operator* in quantum mechanics has the similar form:

$$\hat{\mathcal{H}} = \frac{\hat{p}^2}{2m} + V(\hat{x}) \qquad \text{(A-2)}$$

but for the fact that $\hat{p}$ is the momentum *operator* whose action needs to be defined. As an example, for a quantum harmonic oscillator of angular frequency ω, the Hamiltonian operator is

$$\hat{\mathcal{H}} = \frac{\hat{p}^2}{2m} + \frac{m}{2} \cdot \omega^2 \hat{x}^2$$

Dirac (1958) noted that, "there is a limit to the fineness of our powers of observation and the smallness of the accompanying disturbance—a limit which is inherent in the nature of things, and can never be surpassed by improved technique."

A rigorous statement of this idea is known as the *Heisenberg*, or "*Uncertainty Principle*": The uncertainty in the position, Δx and the uncertainty in the momentum, Δp, of any quantum system cannot be smaller than the Planck's constant (Planck, 1901) divided by 2π (written as $\hbar$):

$$\boxed{\Delta x \cdot \Delta p \geq \hbar} \qquad \text{(A-3)}$$

For the position to be precisely known, the quantum system must be subjected to a large disturbance that increases accordingly the uncertainty of the momentum, Δp, of the system. The uncertainty principle, as stated in equation (A-3), indicates that the position and momentum are complementary variables, or that there is a direct connection between the mutual disturbances involved in the measurements of position and momentum (*Planck's constant*) whose minimum product must be the *unit of action*, that is, $\hbar$. The algebraic form of this statement is known as the commutation relation between the position and momentum operators:

$$\hat{x} \cdot \hat{p} - \hat{p} \cdot \hat{x} \equiv [\hat{x}, \hat{p}] = i\hbar \tag{A-4}$$

It is said that the "position and momentum operators *commute*," and $[\hat{x}, \hat{p}]$ is called the *commutator* of $\hat{x}$ and $\hat{p}$. In more general terms, the following postulate, which is an extension of the commutation relation (A-3), applies:

The Complementarity Principle. Any position observable, $\hat{x}$, and its corresponding generalized momentum, $\hat{\pi}$ are complementary, in the sense that the operators $\hat{x}$ and $\hat{\pi}$ satisfy a commutation relation analogous to (A-4):

$$\boxed{[\hat{x}, \hat{\pi}] = i\hbar} \tag{A-5}$$

The commutation relation (A-4) becomes transparent when expressed in terms of the action of the operators $\hat{x}$ and $\hat{p}$ on any function $\Psi(x)$. For example, the action of the operator $\hat{x}$ on any function $\Psi(x)$ yields a new function $x\Psi(x)$. If one considers also the operator $\partial/\partial x$, then the commutator $[\hat{x}, \partial/\partial x]$ is equal to -1 because for any function $\Psi(x)$:

$$\left[\hat{x}, \frac{\hat{\partial}}{\partial x}\right] \Psi(x) = \left(x \frac{\partial}{\partial x} - \frac{\partial}{\partial x} x\right) \Psi(x)$$

$$= \left(x \frac{\partial}{\partial x} - 1 - x \frac{\partial}{\partial x}\right) \Psi(x) = -1 \cdot \Psi(x)$$

Since $\left[\hat{x}, \dfrac{\hat{\partial}}{\partial x}\right] = -1$, then

$$\left[\hat{x}, -i\hbar \frac{\hat{\partial}}{\partial x}\right] = i\hbar \tag{A-6}$$

Comparing equation (A-6) with equation (A-4) gives

$$\boxed{\hat{p} = -i\hbar \frac{\hat{\partial}}{\partial x}} \tag{A-7}$$

Combining equation (A-6) with the natural representation

$$\boxed{\hat{x} \longrightarrow x}$$

(A-8)

for the position operator, results in what is known as the *Schrödinger representation*.

Finally, combining the Schrödinger representation (A-7 and A-8) with the complementarity and correspondence (A-2) principles gives the explicit form of the *energy operator*, or the *Hamiltonian operator* of a quantum system:

$$\hat{H}(\hat{x}, \hat{p}) = \hat{H}\left(x, -i\hbar \frac{\partial}{\partial x}\right)$$

(A-9)

The action of the Hamiltonian operator on the wavefunctions $\Psi_E(x)$ of a quantum system determines the possible energy values (eigenvalues) of the system:

$$\boxed{\hat{H}\left(\hat{x}, -i\hbar \frac{\partial}{\partial x}\right) \Psi_n^E(x) = E_n \Psi_n^E(x)}$$

(A-10)

Equation (A-10) is known as the *Schrödinger equation* (see also eq. (3-2)) and it plays a key role in quantum mechanics. As an example, consider a particle that moves "freely" in a potential $V(x, y, z)$. Using equations (A-2) and (A-9), the corresponding form of the Schrödinger equation for such a particle is obtained:

$$\left[-\frac{\hbar}{2m} \cdot \left(\frac{\partial^2}{\partial x^2} + \frac{\partial^2}{\partial y^2} + \frac{\partial^2}{\partial z^2}\right) + V(x, y, z)\right]$$
$$\cdot \Psi_n^E(x, y, z) = E_n \Psi_n^E(x, y, z)$$

(A-11)

Equation (A-11) should be solved subject to the condition that $\Psi_E^n(x, y, z)$ is finite everywhere. The operator

$$\frac{\partial^2}{\partial x^2} + \frac{\partial^2}{\partial y^2} + \frac{\partial^2}{\partial z^2}$$

is usually written in shorthand as the Laplace operator, $\Delta \equiv \nabla^2$, where ∇ is the "nabla" operator.

References

Bohr, N. 1913. *Phil. Mag.* **26**:476–484.

de Broglie, L. 1925. *Ann. Physik* **3**:22–28.

Coulson, C. A. 1982. *The Shape and Structure of Molecules*, 2d ed., revised by R. McWeeney. New York: Oxford University Press.

Davisson, C., and L. Germer. 1927. *Nature* **119**:558–560.

Dirac, P. A. M. 1958. *The Principles of Quantum Mechanics*, 4th ed. New York: Oxford University Press.

Kittel, C. 1963. *Quantum Theory of Solids*, chap. 5. New York: Wiley.

Mathews, P. T., F. R. S. 1968. *Introduction to Quantum Mechanics*, 2d ed. London and New York: McGraw-Hill.

Parker, S. P., ed. 1983. *McGraw-Hill Encyclopedia of Chemistry.* New York: McGraw-Hill.

Pauling, L. 1960. *The Nature of the Chemical Bond*, 3d ed. Ithaca, N.Y.: Cornell University Press.

Planck, M. 1901. *Ann. Physik* **4**:553.

Pullman, B., and A. Pullman. 1963. *Quantum Biochemistry*. New York: Wiley-Interscience.

Pullman, B., and A. Pullman. 1969. Quantum chemistry of polynucleotides and nucleic acids. *Prog. Nucleic Acid Res.* **9**:327.

Schaefer, H. F., III. 1972. *The Electronic Structure of Atoms and Molecules*. Reading, Mass.: Addison-Wesley.

Streiwieser, A., Jr., and P. H. Owens. 1973. *Orbital and Electron Density Diagrams*. New York: Macmillan.

Tinoco, I., Jr., K. Sauer, and J. S. Wang. 1978. *Physical Chemistry*, 2d ed. Englewood Cliffs, N. J.: Prentice-Hall.

4

Molecular Structures and Chemical Bonding

The aim of the following ten sections is to discuss briefly some of the basic modern concepts used in the description of molecular structures. The *chemical structures* of the major components of food—*water, carbohydrates, lipids, and proteins*—are presented with selected samples and illustrations.

ATOMIC STRUCTURE

Atoms are about 10^{-8} cm in "diameter" and consist of a positively charged nucleus surrounded by a number of electrons. Both change with the type of atom, hydrogen being the simplest, with one proton and one electron. Most of the mass of the atom is concentrated in the nucleus, which is about 10^{-13} cm in "diameter" and is composed of protons and neutrons. Protons have a positive charge that is numerically equal to that of the electron, but neutrons have no charge. Atoms are electrically neutral, which means the number of electrons surrounding the nucleus is equal to the number of protons in the nucleus, and this number, the *atomic number* of the element, represents its position in the periodic table. The following three statements summarize in a simplified form the concepts developed in Chapter 3 that are needed in this chapter.

The chemical reactivity of an element depends on the electronic structure of its atoms. The modern theory of the electronic structure of atoms has developed from the Bohr theory, which pictured electrons as occupying fixed orbits of definite energy surrounding the nucleus. With the development of quantum mechanics the orbits of the Bohr model were replaced by *orbitals*, whose shape represent the probability of finding an electron with a particular energy. Orbitals are characterized by two quantum numbers, and in order of increasing energy they are represented as 1*s*, 2*s*, 2*p*, 3*p*, 4*s*, 3*d*, 4*p*, 5*s*, 4*d*, 5*p*, 6*s*, ... The

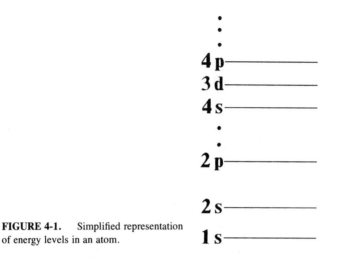

FIGURE 4-1.   Simplified representation of energy levels in an atom.

numbers represent one quantum number and the lowercase letters represent the other quantum number (where s, p, d, correspond to the numerical values 1, 2, 3, respectively). The corresponding energy levels are represented in Figure 4-1. Details of the calculation of these energy levels are given in Chapter 3.

The s-orbitals are spherically symmetrical, but the others are of more complex shape. The p-orbitals, for example, are elongated along three perpendicular axes, as shown in Figure 4-2a, and can be represented as p_x, p_y, and p_z. An electron can be represented as either "spinning" in a clockwise or an anticlockwise direction, and according to the *Pauli exclusion principle*, only one electron of a given spin can exist in a given orbital. There are five different d-orbitals, so that a maximum of ten electrons can be accommodated in them (Fig. 4-2b).

The various elements in the periodic table contain different numbers of electrons, and these electrons are distributed among the various orbitals. Orbitals of lower energy are the first to be filled, followed by those of higher energy. Particularly stable electronic arrangements are obtained when certain orbitals are filled with electrons. These represent electronic arrangements in helium, neon, argon, krypton, xenon, and radon, elements that are chemically very inert. For other elements the various orbitals are not completely filled with electrons, and this is what determines their chemical reactivity and the kind of bonds that they form.

CHEMICAL BONDS

Ionic Bonds

One way of forming a *bond* between two atoms is by means of *electron transfer*. If an atom gains or loses an electron, it is no longer electrically neutral, and it

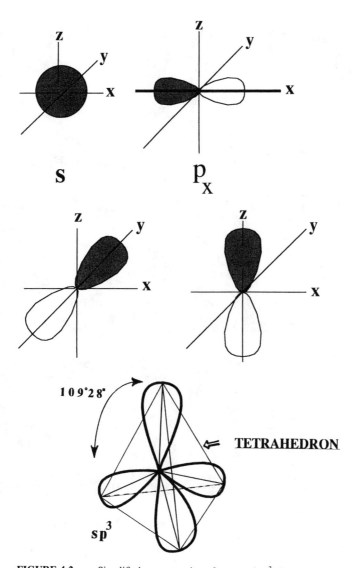

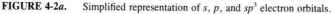

FIGURE 4-2a.   Simplified representation of s, p, and sp^3 electron orbitals.

is then called an *ion*. If an atom gains an electron at the expense of another atom, then an electrostatic attraction exists between the two ions. This is referred to as an *ionic bond*. One example of a structure held together by ionic bonds is a sodium chloride crystal, where the sodium ion is positively charged and the chlorine ion is negatively charged. The transfer of one electron from the neutral sodium atom to the chlorine atom results in each ion attaining the

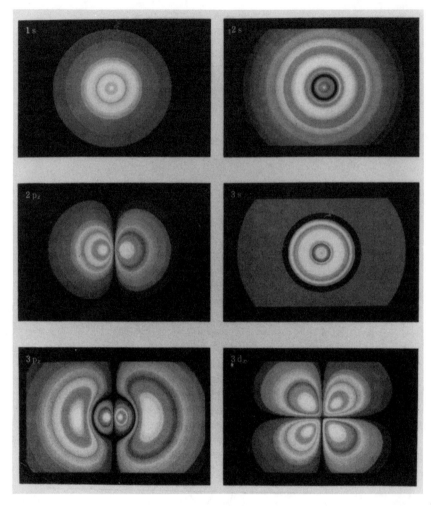

FIGURE 4-2b.   Computer calculations of electron orbitals in hydrogen (from Tokita, 1987, with permission).

stable electronic configuration of the rare gases, that is, the sodium ion attains the helium arrangement and the chlorine ion that of argon. In order to represent the ionic nature of sodium chloride, its structure is represented as Na^+Cl^-. Ionic bonds have no directional characteristics.

Covalent Bonds

An alternative method of attaining a stable electronic arrangement is by *electron sharing*. The atoms that share electrons are attracted to each other, and a *co-*

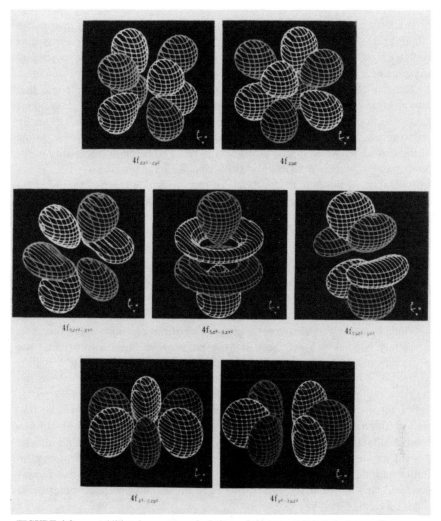

FIGURE 4-2c.   Additional computer calculations of electron orbitals in hydrogen (from Tokita, 1987, with permission).

valent bond is formed between them. The atomic orbitals of two atoms that are covalently bonded interact, resulting in molecular orbitals. Thus, two hydrogen atoms attain stable electronic arrangements when they share their two electrons and a molecular orbital is formed by overlap of the atomic orbitals.

A covalent bond that is formed by the sharing of a pair of electrons is called a *single bond*. When more than one electron is required to give a stable electronic configuration, then more than one single bond can be formed; that is, a double bond involving two pairs of electrons, or even a triple bond involving three pairs of electrons. When a covalent bond is formed between different types of atoms, the shared electrons may be attracted more to one atom than to the other. This gives rise to a separation of charges along the bond, which thus acquires a *polar* character. The more an atom attracts electrons, the more *electronegative* it is said to be.

The fact that most atomic orbitals are not spherically symmetric and that they overlap in the formation of covalent bonds explains why covalent bonds have *directional characteristics*. In addition to the directions accounted for by the atomic orbitals referred to already, other directions are explained in terms of the *hybridization* of atomic orbitals, a process whereby different atomic orbitals blend together to form hybrid orbitals. An example of this effect is the hybridization of the $2s$ and $2p$ orbitals of carbon, which results in four hybrid sp^3 orbitals tetrahedrally arranged around the carbon atom (Fig. 4-2a).

Hydrogen Bonds

A *hydrogen bond* is formed when a hydrogen atom that is covalently bonded to an atom, A, is also attracted to another atom, B. The two atoms A and B are thus linked together through the hydrogen atom.

Hydrogen bonds are formed between highly electronegative atoms such as fluorine, oxygen, and nitrogen. When one of these electronegative atoms is covalently bonded to a hydrogen atom, a polar group is formed, because the electrons are partly drawn away from the hydrogen atom, leaving it with a partial positive charge. The hydrogen atom will thus be attracted toward the negative region of an adjacent electronegative atom. In biomolecules, polar —NH and —OH groups occur very often, and partake in hydrogen bond formation. In contrast to these groups, nonpolar CH groups do not form hydrogen bonds. Although the strength of a hydrogen bond is less than one-tenth that of a covalent bond, hydrogen bonds play a very important part in determining the conformation of biomolecules, both by intra- and intermolecular bond formation, and also by the formation of hydrogen bonds between polar groups and water molecules.

A hydrogen bond is represented by a continuous line for the covalent bond to the hydrogen atom and a dotted line to the second atom. Thus, for example, a hydrogen bond between a nitrogen and an oxygen atom is represented as

$$\diagdown N \text{———} H \ldots O {=\!\!=} C \diagup \qquad \text{(F-4-1)}$$

The length of a hydrogen bond is the distance between the nonhydrogen atoms, and the values are from about 2.8 to 3.0 Å. A more detailed, quantitative presentation of molecular interactions through hydrogen bonds is postponed until Chapter 6.

Bond Lengths

The analysis of X-ray diffraction, electron diffraction, and spectroscopic data have allowed the determination of *interatomic distances*. These results have, in turn, enabled us to determine covalent bond radii, from which the length of covalent bonds between atoms can be predicted with a fair degree of accuracy.

Deviations from the predicted values may arise due to partial ionic characteristics when bonds are formed between different atoms, but corrections can be applied in such cases. Other deviations that occur in some structures are explained in terms of "resonance," due to more than one possible electronic configuration in a molecule. For example, in the benzene molecule the C==C bond lengths are 1.39 Å, a value that lies between the single bond value of 1.544 Å and the double bond value of 1.334 Å, and the structure of benzene instead of being represented by

should be represented as

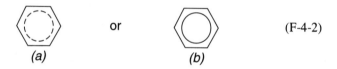

$$\text{(a)} \quad \text{or} \quad \text{(b)} \qquad \text{(F-4-2)}$$

with the inner ring representing *delocalized* electrons in π-molecular orbitals shaped as two doughnuts, above and below the plane of the carbon nuclei. This formation leads to a definition of the bond order of a covalent bond, which is intermediate between that of a single and a double bond.

A second example is the C—N bond length in the peptide bond that occurs in protein molecules. This bond length is found to be 1.32 Å, whereas the C=N bond is 1.28 Å and a C—N bond is 1.48 Å. The bonding electrons participating in the peptide bond are delocalized over oxygen, carbon, and nitrogen, causing these atoms to lie in the same plane, O==C==NH.

Two examples of bond by conversion of combinations of atomic orbitals into molecular orbitals are shown in Figures 4-3 and 4-4 for ethane (single-bond formation) and ethylene (double-bond formation), respectively.

Bond Angles

The directional characteristics of covalent bonds can be explained in terms of the directions of the atomic orbitals. The *p*-orbitals, for example, are mutually

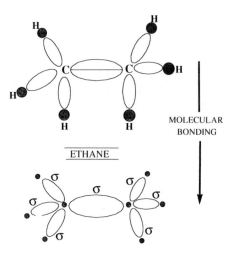

MOLECULAR
BONDING

ETHANE

FIGURE 4-3. Single-bond formation in ethane represented as a combination of atomic orbitals and conversion into (σ) molecular orbitals. Note the cylindrical symmetry of the σ-electron orbital responsible for the single C—C bond.

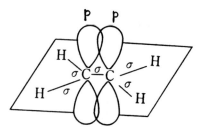

ETHYLENE

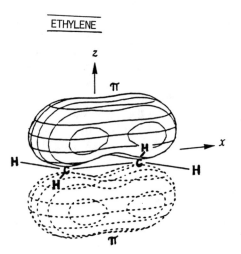

FIGURE 4-4. Double-bond formation in ethylene, represented as a combination of p-atomic orbitals (*top*) and conversion into molecular π-orbitals corresponding to delocalized (π) electrons (*bottom*).

perpendicular, so that bonds formed by p-orbital overlap would be expected to be at right angles to each other. Sulphur atoms form bonds by means of p-orbitals and the angle between the SH bonds in H_2S is $92°$. Deviations from the expected values can occur if the bonds have partial ionic characteristics because the bonded atoms can then repel each other. Thus, in H_2O, where the bonds are formed by p-orbitals of the oxygen atom, the angle between the OH bonds is increased to $\sim 105°$ because of the repulsion between the hydrogen atoms. When molecules pack together in a crystal, steric hindrance between molecules, or between different parts of a molecule, may also cause changes of a few degrees in bond angles.

Other deviations from the expected atomic orbital directions can be described in terms of *hybridization*. Thus, the tetrahedral arrangement of the CH bonds in methane is explained in terms of the hybridization of the three $2p$-orbitals and the single $2s$-orbital of carbon to form four sp^3-orbitals that are tetrahedrally arranged about the carbon atom. The sp^3 hybrid atomic orbitals can be considered intermediates that lead to the formation of four single bonds when they interact with the orbitals of the neighboring atoms. The small instability caused by the formation of the sp^3 hybrids is readily overcome by the additional stability of the four final bonds, as indicated in Figure 4-5.

When carbon–carbon double bonds are formed, it is suggested that hybridization of the $2s$- and two $2p$-orbitals occurs, resulting in three sp^2 hybrid or-

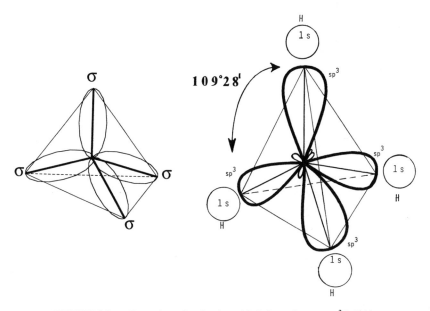

FIGURE 4-5.   Formation of molecular orbitals in methane as sp^3 hybrids.

bitals that are coplanar and at 120° to each other. In the case of carbon–carbon triple bonds the hybridization is believed to be due to the 2s- and one 2p-orbital, which form two sp^1 hybrid orbitals that are colinear.

The most important bond lengths and bond angles encountered in organic molecules and biomolecules are summarized in Figure 4-6.

Rotations about Bonds

In any molecule there is free rotation about single bonds and *no* rotation around double or triple bonds. Delocalized π-electrons, as in benzene or in the peptide bond, also do *not* allow rotation about such bonds. The interatomic distance between the atoms of two specific groups in a molecule that are linked by a single bond restricts the possible relative orientations of such groups, although in principle there is free rotation about the single bonds. The rotation about single bonds causes the large number of possible conformations of molecules containing such bonds.

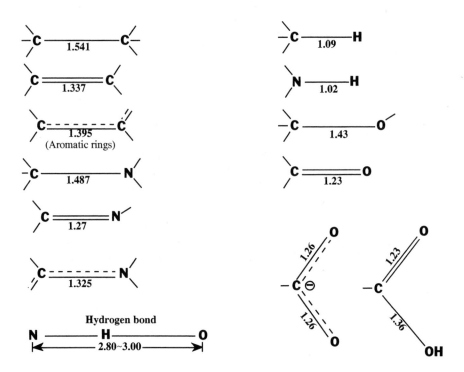

FIGURE 4-6.   Important bond lengths and bond angles in biomolecules (all lengths are in angstrom units).

Van der Waals Interactions

As molecules approach each other very closely, or as different groups of the same molecules come into proximity without forming bonds, they attract each other relatively weakly. Nonbonded atoms continue to approach each other until the outer electron orbitals of such atoms interpenetrate and the resulting repulsion balances out the weak attractive Van der Waals forces. The closest distance of approach of nonbonded atoms at equilibrium defines the Van der Waals radii of atoms. Values of Van der Waals radii for the atoms present in biomolecules are compared in Table 4-1 with the single-bond and double-bond covalent radii. A quantitative treatment of these interactions is presented in Chapter 6.

Molecular Asymmetry and Optical Activity

Molecules with asymmetric structures, that is, those that do *not* possess a mirror (plane) symmetry, cause a rotation of the plane of polarization of plane-polarized light. An example of such an asymmetric, optically active molecule is shown in Figure 4-7*a*. If the rotation is anticlockwise, the molecule is called *levorotatory*; when the rotation is clockwise, the molecule is said to be *dextrorotatory*.

Two asymmetric molecules that are mirror images of each other are called *enantiomorphous*, and they rotate the plane of polarization of light in opposite directions. An enantiomorphic pair is illustrated in Figure 4-7*b*. For example, all amino acids, except glycine, occur in enantiomorphous configurations, referred to as the L- and D-configurations. Only the L–amino acids are present in proteins.

TABLE 4-1   Van der Waals Radii

Atom or Compound	Radii (Å)
H	1.2
F	1.4
Cl	1.8
Br	2
N	1.5
O	1.4
S	1.9
P	1.9
CH_3 (methyl)	2
CH_2 (methylene)	2
One-half thickness of aromatic ring	1.85

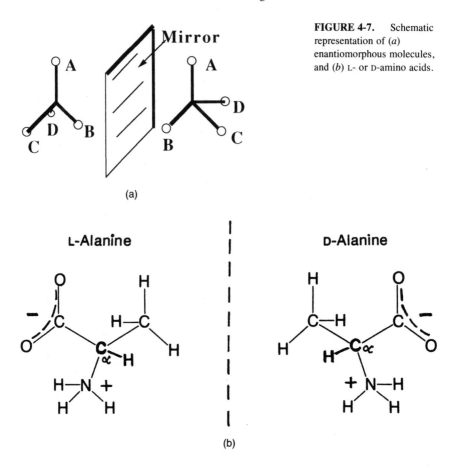

(a)

(b)

Molecular Models

The simplest molecular model uses stiff wires to represent bonds (Fig. 4-8a), and allows for free rotation about single bonds. Such a model is used when accurate atom coordinates are required.

Ball-and-stick models are small colored balls to represent various atoms and wires to represent the bonds between them (Fig. 4-8b). These models are often employed for educational purposes and to build crystallographic models of the simpler structures, such as inorganic salts, diamond, and "hexagonal" ice. Accurate angles and reproducible lattice structures can be obtained with such models.

The third type of molecular model (Fig. 4-8c) is a space-filling (or CPK) model in which the atoms are represented by spheres (or parts of spheres) whose radii are proportional to the Van der Waals radii (Table 4-1). The connections

CH$_4$ C$_6$H$_6$

(a)

(b)

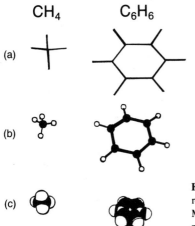

(c)

FIGURE 4-8.   Three types of molecular models representing methane (CH$_4$) and benzene (C$_6$H$_6$). Models: (*a*) Wire. (*b*) Stick-and-ball. (*c*) CPK (close-packed).

between the spheres allow for rotation about single bonds. Whereas the CPK model represents molecular shapes, steric effects, and Van der Waals interactions quite well, it does not allow one to see inside large, complex molecules. The CPK models are also somewhat easier to assemble, but do not allow accurate measurements of bond angles to be made.

STRUCTURE OF THE WATER MOLECULE AND ITS MOLECULAR IONS

Water is a major component of most natural foods and of all living organisms; hence, its importance in the food industry, food science, and the life sciences.

Molecular Structure of Water

Water is a nonlinear, polar molecule containing an oxygen atom bonded to two hydrogen atoms by covalent bonds. Because the electrons are tightly held by the oxygen in the water molecule and the electron density is highest near the oxygen nucleus (O), the charge distribution in a water molecule has a center of negative charge close to the oxygen atom, and centers of positive charge on the two hydrogen atoms (which can be regarded as ''embedded'' in the molecular orbitals (MOs) surrounding the oxygen).

The molecular geometry of water is represented as a space-filling model in Figure 4-9*a*. The OH bond distance is about 0.957 Å, and the HOH bond angle is 104.5°, resulting in an overall diameter of the water molecule of about 3.3 Å. The Van der Waals radius of liquid water is, however, considered to be 1.38 Å. In a first approximation, the charge distribution of the water molecule

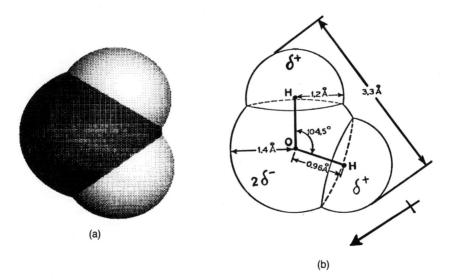

(a)

(b)

FIGURE 4-9.   (*a*) CPK molecular model of the water molecule. (*b*) Direction of the dipole moment of water.

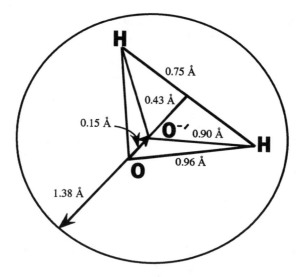

FIGURE 4-10.   Charge distribution in a water molecule according to Bernal and Fowler (O is the center of the oxygen atom and O^- that of the negative charge).

is represented by an electric dipole moment, as indicated in Figure 4-9b. More elaborate models of the charge distribution (Fig. 4-10), which represent the water molecule as an electric quadrupole, were considered.

Coulson and Eisenberg proposed a more accurate representation of the charge distribution in the single water molecule based on the point dipole, quadrupole, and octopole located at the oxygen atom (cited by Conway, 1981). The calculated electric dipole moment with this model is 2.42 debye (D) at 298 K (25°C), whereas the value for the water vapor is 1.84 $\pm$ 0.02 D.

Quantum mechanics, however, provides the best representation of the bonding and electron density distribution in the water molecule. Coulson applied the molecular orbital (MO) theory to water and considered the two OM bonds in the water molecule as MOs formed from one of the 2p orbitals of the oxygen (which has the electron configuration [1s^2, 2s^2, 2p^4]) mixed with the 1s orbital of the hydrogen atom. There would be four electrons (1s^2, 2s^2) in spherical electron orbitals centered on the oxygen, whereas two paired electrons would be located in a 2p_x orbital, with its axis normal to the HOH plane of the water molecule. In this simple representation two electrons, each of which would be placed in the 2p_y and 2p_z bonding orbitals, would remain at the oxygen and would overlap with the two 1s electron orbitals centered on the two hydrogen atoms; these two orbitals would then represent the two OH bonds in the water molecule (Fig. 4-11). This model is, however, oversimplified, since it predicts an HOH bond angle of 90° (Fig. 4-11, bottom), instead of the observed value of 104.5° (Fig. 4-9). For the oxygen to bind covalently to the two hydrogens, there must be sufficient electron charge-density between the oxygen and hydrogen nuclei to counteract the electrostatic repulsion between them. With improved wavefunctions and by considering sp^3 *hybridization* of the participating electron orbitals, the calculated electron density of the water molecule becomes more realistic (Fig. 4-12). The three-dimensional structure of the water molecule in this representation is quasi-tetrahedral; that is, the two bond-hybrids, as well as the two lone-pair orbitals, are arranged symmetrically, almost tetrahedrally, around the oxygen atom. Schematic diagrams of this arrangement are presented in Figure 4-13a and b.

Vibrational Modes of Free Water Molecules

Because H_2O is a nonlinear triatomic molecule, there are three normal modes of vibration (in general, [3 × r_{atoms} − 6]modes), which are shown in Figure 4-14. These modes are symmetric stretching, asymmetric stretching, and bending, as indicated in Figure 4-14. Such vibrations and their combinations lead to fairly complex infrared (IR) absorption spectra of water molecules in the gas phase. By employing liquid heavy water (D_2O) significant simplifications of the IR spectra occur, thus allowing the resolution of many absorption bands of the

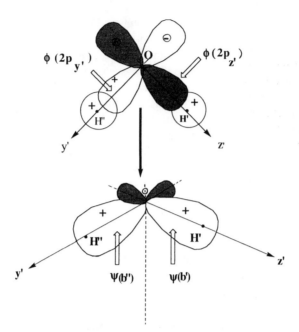

FIGURE 4-11.   Representation of the molecular orbitals in the water molecule, indicating the bond formation as a result of atomic orbital combination.

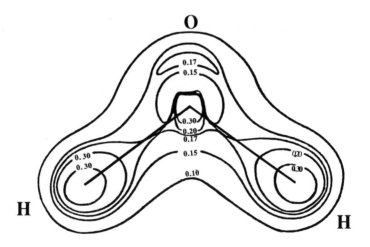

FIGURE 4-12.   Electron density contour diagram of a water molecule (according to Bader and Jones, 1978, cited by Conway, 1981).

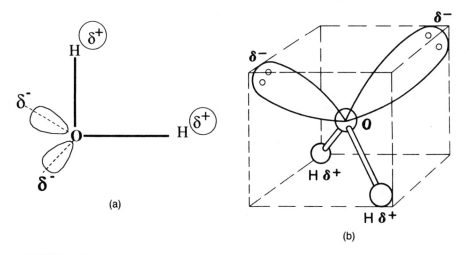

FIGURE 4-13.   Quasi-tetrahedral structure of the water molecule (*a*) involving two bond-hybrids and two lone-pair orbitals, symmetrically arranged around the oxygen (*b*).

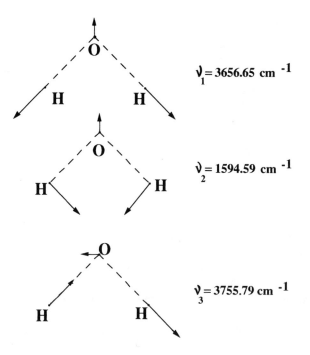

FIGURE 4-14.   Diagram showing the three normal vibrational modes of a free water molecule and defining ν_1, ν_2, and ν_3.

liquid. The IR and Raman absorption frequencies of the maximum vibrational band of water molecules allow the test of selected potential energy functions, ΔU of the water molecules, and are therefore of basic interest in molecular dynamics studies of water.

Molecular Volume of Water

The volume occupied by a water molecule will depend on its environment. Several such volume estimates were derived under different conditions:

1. From the molecular dimensions, V_{calc} (H_2O) = 7.5 cm^3/mol
2. Close-packed water, V_{calc} (H_2O) = 9.7 cm^3/mol
3. Liquid water at 277 K (4°C), V (H_2O) = 18 cm^3/mol
4. Rotational volume of a single H_2O molecule, 22.5 cm^3/mol

Depending on the system under consideration, one may choose the appropriate value from this list.

The discussion of other physical properties of water is postponed until Chapter 6, where the local structure of liquid water and hydrogen bonding between water molecules are discussed in detail.

Molecular Ions of Water and Ionization of Water

At neutral pH and 25°C one water molecule in every 10 million becomes ionized through the bonding of one of its protons to another water molecule, as shown in reaction (R-4-1).

$$
\begin{array}{c}
H \\
\diagdown \\
\diagup \quad O\ldots H{-}O \\
H
\end{array}
\quad
\begin{array}{c}
H \\
\diagdown \\
\end{array}
\rightleftharpoons
\begin{array}{c}
H \\
\diagdown \oplus \\
O{-}H + OH^- \\
\diagup \\
H
\end{array}
\qquad (R\text{-}4\text{-}1)
$$

Two molecular ions are formed as a result of the ionization reaction of water: the *hydronium* ion, $(H_3O)^+$, and the *hydroxyl* ion, $(OH)^-$. The charges belong to the entire molecular group rather than to a single atom, although for simplicity the brackets are omitted and the ions routinely denoted as H_3O^+ and OH^-. Mass spectrometry of the hydrated hydronium in the gas phase identified the stable species $(H_9O_4)^+$, which is in fact $(H_3O)^+ \cdot 3H_2O$. This finding suggests that hydronium is hydrated by three water molecules, as shown in Figure 4-15a. The molecular geometry of $(H_3O)^+$ is shown in Figure 4-15b, where the oxygen is situated slightly higher than the equilateral triangle of its bonded hydrogens. All three hydrogens are therefore *chemically equivalent*, and are located at an equal distance $r_{OH} \simeq 0.98$ Å from the oxygen (as found by nuclear magnetic resonance spectroscopy). The bond structure therefore must be

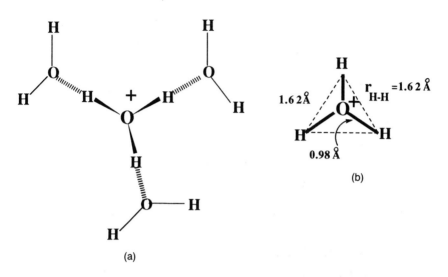

FIGURE 4-15. (*a*) Hydration of the hydronium (hydroxonium) ion with three water molecules that are hydrogen bonded symmetrically to the ion. (*b*) Molecular geometry of the hydronium $(H_3O)^+$ ion.

the result of hybridization with delocalized electrons along the three bond axes. The distance between the hydrogen atoms, r_{HH}, in $(H_3O)^+$ was determined by neutron diffraction (ND), and also independently by nuclear magnetic resonance (NMR). From ND measurements the angle HOH was found to be 109°. The hydronium ion is formed first because of the high mobility and reactivity of the proton (H^+) species both in liquids and in solids (H^+ has the smallest possible cation radius, since *no* electrons are present). The positive charge of $(H_3O)^+$ is centered very close to the oxygen atom as a result of the *delocalization* of the bonding electrons in its molecular orbitals.

The hydronium ion is especially important, since addition of acids to water always results in a corresponding increase in the hydronium concentration. Often, the $(H_3O)^+$ concentration is equated with the H^+ concentration, although each H^+ formed in water is rapidly converted to $(H_3O)^+$ and becomes further hydrated with three water molecules. The concentration of the hydronium ions in an aqueous solution is conventionally used to define the pH scale

$$pH = -\log_{10} [H_3O]^+ = -\log_{10} [H^+] \qquad (4\text{-}1)$$

Thus, for liquid water at 25°C, with a dissociation constant of 10^{-14} mol/l, the "neutral" pH value is $7.00 \equiv -\log_{10} (10^{-7})$, since $[H^+] = [OH^-]$ and

$k_{water} = [H^+] \cdot [OH^-] = 1.00 \times 10^{-14}$. The pH range is from 0 to 14, or a range of $(H_3O)^+$ activity from 1 to 10^{-14} mol/l.

The counterion of the hydronium in reaction (R-4-1) is the hydroxyl OH^-, which has a hydration number of 4 and an ionic radius of about 1.2 Å. The reverse reaction to the ionization of water results in the neutralization of $(H_3O)^+$ by OH^- and the formation of two neutral water molecules from these ions (see reaction (R-4-1)).

CARBOHYDRATES

Carbohydrates have wide applications in food processing. Their properties and behavior in complex mixtures and solutions or gels are therefore practically very important.

Definition: *Carbohydrates* are polyhydroxy aldoses and ketoses, or substances that yield such compounds upon hydrolysis, and are normally divided into three major groups on the basis of their molecular weight: monosaccharides, oligosaccharides, and polysaccharides.

Monosaccharides

Monosaccharides are low molecular weight carbohydrates that cannot be hydrolyzed to smaller molecules that could be defined as carbohydrates. Those monosaccharides with an aldehidic structure are called *aldoses*, while those with a ketonic structure are called *ketoses*. The number of oxygen atoms (expressed in Greek) determines the label of the monosaccharide; thus, a *tetrose* has four oxygens, a *pentose* has five, and a *hexose* has six. These simple monosaccharides, which crystallize readily, have a sweet taste, and are very soluble in water, are commonly called *sugars*. In general, carbohydrates are optically active.

The D-monosaccharides have the general chemical formula

$$(F\text{-}4\text{-}3)$$

where R is the carbonyl group. This is written by analogy with $D(+)$-glyceraldehyde,

$$
\begin{array}{c}
O \\
\parallel \\
C-H \\
\vdots\text{-------}\vdots \\
\vdots\,{}^*HC-OH\,\vdots \\
\vdots\quad\vert\quad\vdots \\
\vdots\,CH_2OH\,\vdots
\end{array}
\qquad\text{(F-4-4)}
$$

The position of the OH group at the highest numbered asymmetric carbon atom (*C) determines the D or L character; the highest number is assigned to the C, which is farthest from the carbonyl (R) group. When the OH is located at the right of the *C atom, as in formulas (F-4-3) and (F-4-4), the stereoisomer has the D label. D-sugars are sweet, whereas L-sugars are not. In addition to formula (F-4-3), four other types of structure representations are in use: (1) the Fischer-projection formula, (2) the Fischer–Tollens formula, (3) the Haworth formula, and (4) the conformational ("boat" and "chair") formulas. The Haworth formulas are shown in Figure 4-16a and b. The Fischer–Tollens formula,

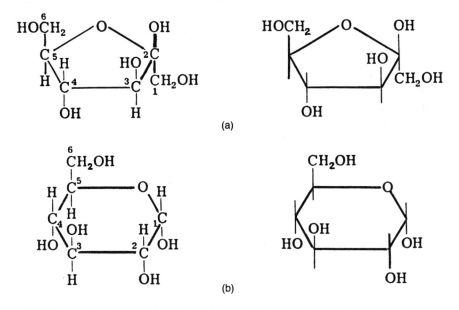

FIGURE 4-16.   Haworth projection formula for 5-member (a) (furan) and 6-member (b) (pyran) sugar rings.

based on the *cyclic hemiacetal*, helps us visualize better the optical activity of the sugars than does the Fischer-projection, but it does not explain satisfactorily the reactions and properties of the sugars. The Haworth formula (Fig. 4-16*a*,*b*) provides a more complete description in terms of 5 (furan) or 6 (pyran) rings of atoms, with one atom being oxygen and the others being carbons. However, *furanose*, for example, is a *strained* ring, whereas *pyranose* is a *strainless* ring; these atoms exist in the "boat" (B) and "chair" (C) forms, shown below:

a = axial bond
e = equatorial bond

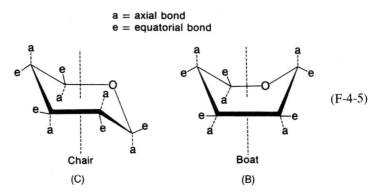

(F-4-5)

Chair   Boat

(C)   (B)

The C atoms 1 and 4 are both above the plane of the other four atoms in the "boat" conformation and on opposite sides of that plane from the other four atoms, whereas the latter are on opposite sides of that plane in the "chair" conformation. Although the conformational formula is the most accurate, the Haworth formula is more often used because of its simplicity.

The Aldoses

In the *cyclic hemiacetal* formulas, such as the Fischer–Tollens formulas for D-glucose, for example,

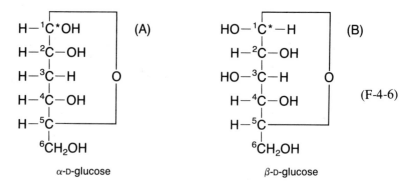

α-D-glucose   β-D-glucose

there is a new asymmetric carbon formed at the 1 position, written $*^1C$ and called the *anomeric carbon*. The two forms of the sugar, (A) and (B) in F-4-6, are called α- and β-anomers, respectively. In the α-anomer, the anomeric OH (linked to the anomeric carbon) is on the same side as the OH that determines the D (or L) character; in the β-anomer, the anomeric OH is on the opposite side of the OH that determines the D (or L) character.

When an aqueous solution of monosaccharide is allowed to stand, the optical activity of the solution changes with time until it reaches a constant value. This process is called *mutarotation*. Furthermore, the Schiff's reagent for aldehydic groups does not react with the sugars, because the structure is not open as in the Fischer-projection formula (Fig. 4-16a), but involves the cyclic hemiacetal formulas (F-4-6). Thus, the mutarotation process can be represented by the reactions

$$\alpha\text{-D-glucose} \rightleftharpoons \text{free aldehyde form} \rightleftharpoons \beta\text{-D-glucose}$$

$$\text{D-glucose (open)} \tag{R-4-2}$$

At equilibrium, the molar fractions of these forms are α-D-glucose 37%, free aldehyde form 1%, and β-D-glucose 62%. The specific rotation, $(\alpha)_{D^0}^2$, is $+113°$ for the α-anomer and $+19°$ for the β-anomer of D-glucose. A solution of D-glucose will attain a specific rotation value of 52.5° at equilibrium, consistent with the preceding values of $(\alpha)_{D^0}^2$ at equilibrium and the molar fractions of the conformers.

All sugars that reduce Fehling's reagent or react with the Benedict solution will exhibit mutarotation. If one blocks the mutarotation by forming, for example, D-methyl glucose, then this sugar will no longer have a reducing ability toward Fehling's reagent. Some ketoses also exhibit mutarotation, but their behavior is more complex than that of aldoses.

With Haworth's formula, the mutarotation of D-glucose can be represented as

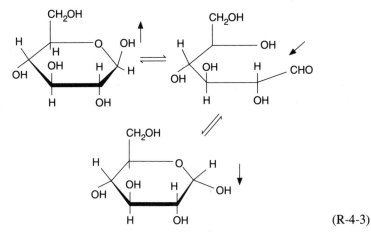

$$\tag{R-4-3}$$

The Ketoses

Whereas the aldose family is built up from glyceraldehyde (formula (F-4-4)), the ketoses are built up from *dihydroxyacetone*

$$
\begin{array}{l}
CH_2OH \\
| \\
C=O \\
| \\
CH_2OH
\end{array}
\qquad (F\text{-}4\text{-}7)
$$

Dihydroxyacetone does *not* contain an asymmetric carbon, however, but beginning with the next larger molecule, erythrulose (formula (F-4-8)), an asymmetric carbon, C^* is present, and optical rotation is observed because of the structure

$$
\begin{array}{l}
CH_2OH \\
| \\
C=O \\
| \\
HC^*OH \\
| \\
CH_2OH
\end{array}
\qquad (F\text{-}4\text{-}8)
$$

The most important number of the ketose series is the fructose molecule, which contains a chain of six carbons:

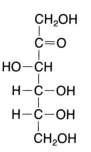

$$
\begin{array}{l}
CH_2OH \\
| \\
C=O \\
| \\
HO-CH \\
| \\
H-C-OH \\
| \\
H-C-OH \\
| \\
CH_2OH
\end{array}
\qquad (F\text{-}4\text{-}9)
$$

Fructose is the *sweetest* of all sugars and is *strongly levorotatory*. It occurs in fruit juices, honey, and several plants. Fructose does *not* exhibit mutarotation.

Fructose, D-glucose, and D-mannose all react with phenylhydrazine $C_6H_5NHNH_2$ to yield the same (phenyl) osazone, implying that fructose has the same configuration (*s*) at atoms C3, C4, and C5 as D-glucose and D-mannose.

The conformation of fructose in solution at equilibrium was recently determined by carbon-13 NMR spectroscopy (Prince et al., 1982). This report discussed the analysis of honey, which is a mixture of about 38% fructose, 31% glucose, 7% maltose, 17% water, and 1% sucrose. The ^{13}C NMR analysis showed that fructose in water is predominantly in the β-D-fructopyranose (formula (F-4-10)) form (67%), whereas the β-D-fructofuranose is only 27% and the α-D-fructofuranose is only about 6%. This is *not*, however, the case in disaccharides (such as sucrose) or in polysaccharides, where this conformation is prevented by the glycosidic link. The metabolic forms of fructose, such as fructose phosphate, are also predominantly in the β-D-fructofuranose form:

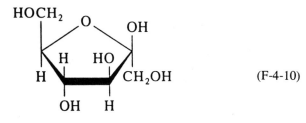

(F-4-10)

β-D-Fructofuranose

Oligosaccharides

By definition *oligosaccharides* consist of two to ten monosaccharide units linked together by single bonds. Disaccharides, which contain only two monosaccharides linked by a glycosidic bond, are the first in this series. The most important disaccharides are sucrose (commonly called "sugar"), lactose, and maltose.

Sucrose
The sucrose molecule is a *nonreducing sugar*, it does not show mutarotation, and does not form osazone with phenylhydrazine. Commercial sources of sucrose are sugar beet and sugar cane, and most ripe fruits are rich in it, while in foods, sucrose is often present at the 0.1% to 2.5% levels.

The formula of sucrose can be written as α-D-glucopyranosyl, 0-1→2-β-D-fructofuranoside, or in the Haworth form

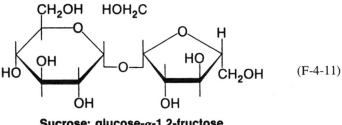

(F-4-11)

Sucrose: glucose-α-1,2-fructose

Sucrose is often chosen as a sweetness "standard," and is assigned a sweetness of 100.0 for a 10% solution.

Lactose

Unlike sucrose, *lactose* is a reducing sugar, undergoes mutarotation, and forms an osazone. The enzyme *lactase* (β-D-galactosidase) hydrolyzes the lactose, yielding equal mole fractions of galactose and glucose, which are therefore its units linked by a β-glycosidic link. The Haworth structural formula of α-lactose, or β-D-galactopyransoyl, 0-1→4-α-D-glycopyranoside is shown below:

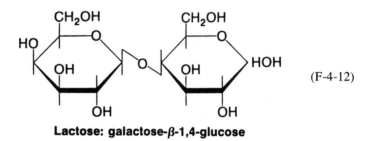

(F-4-12)

Lactose: galactose-β-1,4-glucose

Lactose is the least sweet sugar, and it also has the lowest solubility in water. As the name indicates, lactose is found in the milk of all mammals. It is used in the food industry in baked products and in biscuits. Lactose intolerance in infants (who lack lactase) can be serious, since dehydration may rapidly occur.

There are several trisaccharides occurring naturally in foods. Among them raffinose, which is a nonreducing sugar that contains a fructose, a glucose, and a galactose unit linked together. Although it is nontoxic, raffinose is an undesirable ingredient in foods.

Molecular Structure of Sugars, and Sweetness

The sweetness of sugars is evaluated by sensory methods through tasting panels that compare other sugars with a 10% sucrose solution "standard" ("sweetness" value = 100.0). It has been found that the conformation of the sugar plays a very important role in determining the sensation of sweetness. This role is the basis for a model, or "theory," based on the hypothesis that the conformation of a sweet molecule is such that it fits very closely in a receptor, with which it makes a pair of hydrogen bonds. The groups of the sweet molecules involved in such hydrogen bonds would be OH groups, which are *not* intramolecularly hydrogen bonded. The term "sweetness unit" was coined to designate two such groups, called AH and B, where AH would be the group with the partial negative charge, such as −OH or −NH, while the B group would also have a partial negative charge located no further than 3 Å from the AH group, and may or may not have a hydrogen atom present. A third site, called γ, may also be involved. This site could be a carbon atom with up to three hydrogens attached and would fit a complementary site on the receptor. Schematically, this model assumes that the "sweetness unit" is a triangle, with the

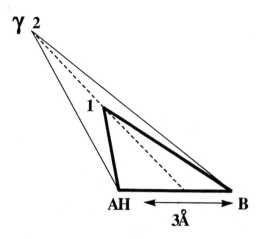

FIGURE 4-17.   Triangle model of the sweetness receptor involving a "sweetness unit" (AH + B) and a third site γ at the apex of the triangle, which acts as an enhancer of the sweetness sensation.

base AH at 3 Å from B and the γ site at the apex, where it acts as an enhancer of the sweetness sensation (Fig. 4-17). This model explains why ethylene glycol, D-amino acids, and D-sugars are sweet, while L-sugars and L-amino acids are not. The exception is L-fructose, which is sweet.

The conformational formulas of glucose and galactose (formulas (F-4-13) and (F-4-14), respectively) are shown with the AH and B groups indicated:

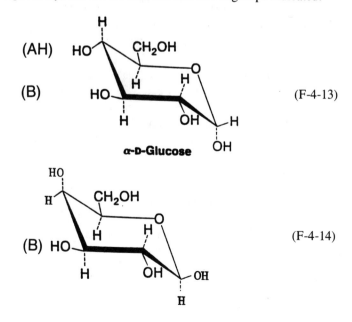

Because the OH group in the top left corner of the galactose molecule can hydrogen bond intramolecularly, galactose is less sweet than glucose. The sweetness remains, however, only because other OH groups can fit somewhat imperfectly into the receptor and hydrogen bond less strongly than the AH and B groups of glucose.

Polysaccharides

Polysaccharides are polymers of ten or more simple monosaccharides linked together by glycosidic linkages. Polysaccharides are sometimes known generically as *glucans*. Naturally occurring polysaccharides usually possess very high degrees of polymerization and contain between 10^2 and 10^6 monosaccharide units. Their degree of polymerization is quite variable, even for a given type. This heterogeneity makes their study more difficult than that of purified substances.

Unlike small carbohydrates, polysaccharides are not sweet, but are almost without taste. The reason for their lack of sweetness can be understood by referring to Figure 4-17.

Classification

1. Homopolysaccharides (or homoglycans). Only *one* monosaccharide is produced by the hydrolysis of homopolysaccharides. Major groups of homopolysaccharides are
   a. amylose, amylopectin (from starch), cellulose, glycogen, and dextran. Their monomer is D-glucose. Glycogen, amylose, and amylopectin are storage polysaccharides.
   b. galactans, whose monomer is D-galactose.
   c. chitin (whose monomer is *N*-acetyl-D-glucosamine) is a structural polysaccharide, along with cellulose and pectins.
2. Heteropolysaccharides. Several monosaccharides are produced by the hydrolysis of heteropolysaccharides. Two important examples of heteropolysaccharides are pectins (which are made of D-galacturonic acid and methyl D-galacturonate), and locust bean gums (which are made of D-mannose and D-galactose). Another is glucomannan, extracted from the konnjak plant grown in several oriental countries.
3. Glycoproteins. In glycoproteins, the carbohydrate units, or oligomers, are attached covalently to a protein. Examples of glycoproteins are membrane glycoproteins, serum glycoproteins, and proteoglycans. Membrane glycoproteins are thought to play important roles in cell-to-cell interactions and cell recognition.

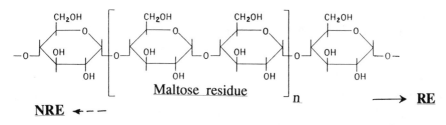

FIGURE 4-18.   Amylose structure (Haworth projection).

Amylose Structure

An amylose structure consists of more than 250 D-glucose units linked "linearly" by the $\alpha(1\rightarrow4)$ glycosidic bond, as shown in Figure 4-18. A partial rotation of all the glycosid bonds results in a helical structure, the interior of which traps iodine, fatty acids, or monoglycerides. Amylose gives a blue color when complexed with iodine in water.

Amylopectin Structure

An amylopectin is a highly branched polysaccharide made of glucose units, whose branching points are $\alpha(1\rightarrow6)$ linkages and whose "linear" branches are made of $\alpha(1\rightarrow4)$-linked D-glucose units (similar to amylose), as shown in Figure 4-19. Amylopectins have between 10^5 and 10^6 glycose units and corresponding molecular weights between 1.6×10^7 and 1.6×10^8. Amylopectins give a *purple color* with iodine in water.

Amylose and amylopectins are the major components of *starch granules*. Upon standing, amylopectins can form gels (retrogradation) by weak intermolecular associations such as H-bonding. The retrograded amylopectin fraction

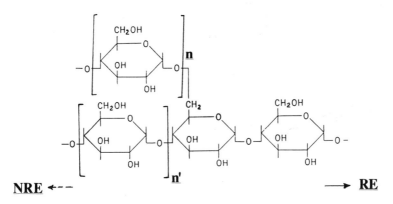

FIGURE 4-19.   Amylopectin structure (Haworth projection).

is redispersed by heating above 50°C, whereas the retrograded amylose will not redisperse. In the case of amylose, these associations are more numerous and stronger H-bonds are formed between the amylose molecules.

Glycogen Structure
Glycogen has a similar structure to amylopectin, only it has more numerous and frequent branches. The average number of glucose units per linear segment is between 8 and 12 for glycogen, and between 24 and 32 for cereal starch amylopectin. Glycogen molecules form clusters, aggregated in large granules that are *noncrystalline* (unlike starch from plants, which is mostly crystalline). This glycogen structure is well-suited for its role as the main storage form of carbohydrates in the liver and muscles. The glycogen in tissues is well-hydrated and forms swollen granules that facilitate rapid enzymatic cleavage, thus making D-glucose readily available as needed.

Cellulose Structure
Cellulose is a linear polysaccharide made of D-glucose units joined by $\beta(1\rightarrow4)$ linkages, and its partial hydrolysis yields a reducing disaccharide called *cellobiose*. The generic conformational formula of cellulose is shown in Figure 4-20. The molecular weight of cellulose varies between 50,000 and 2.5×10^6, corresponding to a number of 300 to 15,000 glucose units. Cellulose is the major structural component of plants (especially cell walls).

Because of their conformation (Fig. 4-20), cellulose molecules readily associate in bundles of long parallel chains. The $\beta(1\rightarrow4)$ linkages determine the conformation shown in Figure 4-20, which is significantly different from the amylose helix permitted by the $\alpha(1\rightarrow4)$ linkages.

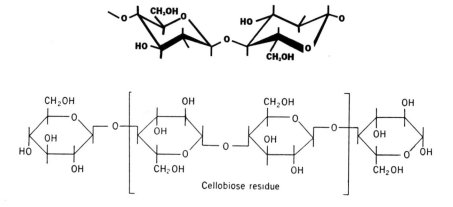

FIGURE 4-20.   Cellulose structure (conformational formula).

Structure of Pectic Substances

Pectic acids are polygalacturonic acids. The occasional substitution of methoxyl groups instead of some carboxyls results in pectins whose structure is illustrated in Figure 4-21. The methoxyl content of pectins varies from low (2% to 6%) to high (7% to 16.32%). The gelation properties of pectins vary with the methoxyl content, with the completely methoxylated pectins forming gels when sugar is added, whereas the low methoxyl pectins will not gel unless polyvalent ions (e.g., Ca^{+2}) are present. There are changes in the structure of pectic substances during ripening and postharvest storage of fruits and vegetables. That is, the enzymatic cleavage of protopectin causes the loss of firmness in fruits and vegetables during ripening and storage.

Gums

All gums are heteropolysaccharides with relatively complex structures, which are still being investigated. Gums are commonly used as food additives, and the most important ones are guar(guaran), locust bean-, arabic-, gums, agar, carrageenans, and alginates. Most gums have either β-D-mannose or α-D-galactose as one of the subunits. Gums generally have a stabilizing effect on food dispersions, due to their strong hydration and their "thickening" action (apparent viscosity increase) in aqueous suspensions. Some gums, such as agarose, kappa-carrageenan, and alginate, have good gelling properties due to their ability to cross-link at sufficiently high concentrations. Arabic gum, on the other hand, is very highly branched, and is recognized as one of the best materials for flavor encapsulation in foods (although relatively costly).

Acid Mucopolysaccharides

Acid mucopolysaccharides (MPS's) are heteropolysaccharides consisting of repeating units of hexuronic acid and hexosamine (Fig. 4-22). In keratan sulfate, however, hexuronic acid is replaced by galactose. Also, the acetyl and sulfate groups occur in varying amounts, especially in the hexosamine components. All MPS's are polyanions, and the negatively charged groups are the carboxyl and sulfate groups. Mucopolysaccharides are major components of the cell

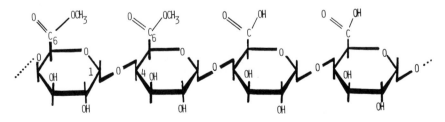

FIGURE 4-21.   The chemical structure of pectins.

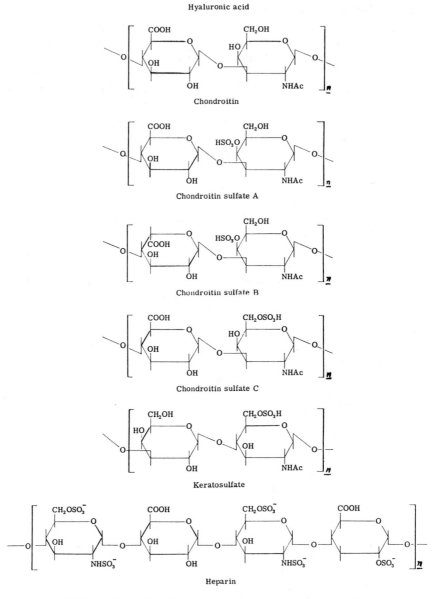

FIGURE 4-22.   Chemical structures of several mucopolysaccharides.

walls, acting as a protecting coating at the surface of some cells. They are also present in the intercellular spaces between cells in tissues, acting as spacers. Mucopolysaccharides give flexibility to connective tissues, and dermatan sulfate is found in the skin and heart valves and aorta.

LIPIDS

From a structural standpoint, lipids can be divided into two categories: (1) simple lipids and (2) complex lipids.

Simple Lipids

Simple lipids are neutral glycerides, esters of fatty alcohols, and include the lipid compounds called *derived* lipids, which are formed from the latter by acid or alkaline hydrolysis. This group of lipids includes fatty acids, fatty alcohols, and glycerides.

Fatty Acids

There are two main groups of fatty acids, *saturated* and *unsaturated*. Their nomenclature is based on that of the corresponding parent hydrocarbons. For example, when the parent hydrocarbon is hexane, the corresponding fatty acid is the *hexaenoic acid*, whereas *octanoic acid*, $CH_3(CH_2)_6COOH$, has octane as the parent hydrocarbon. If the fatty acid has one double bond (such as the one formed from the hydrocarbon octene), it is called *moneonic* (such as the octanoic acid). If two double bonds are present, the fatty acids are called *dienoic*; for example, the octadecadienoic acid has 18 carbons and two double bonds. In general, one can encounter trienoic (3), tetraenoic (4), pentaenoic (5), and hexaenoic (6) acids (with the corresponding number of double bonds indicated in brackets).

The location of the double bond along the fatty acid chain can be indicated either by a pair of numbers or by a Greek letter (ω or η, for example) and a number. A few examples will help explain these labeling conventions. Oleic acid is $CH_3(CH_2)_7 - {}^{10}CH = {}^9CH(CH_2)_7 - {}^1COOH$, with the "beginning" of the chain at the carboxyl end (carbon C-1); the oleic acid is, therefore, defined as 9,10-octadecenoic acid, or simply 9-octadecenoic acid (or Δ9,10-octadecenoic acid). Since most of the naturally occurring fatty acids do not have conjugated double bonds, a shorthand notation was adopted for them. The terminal methyl (CH_3) group is labeled by ω or η, and the number of steps along the chain from that methyl to the first carbon that has a double bond is written beside the letter ω or η to denote the series of fatty acids; for example, if this number of atoms to the first double-bonded carbon is 7 (starting from the ω or η terminal methyl), then the series of fatty acids is known as the $\omega 7$ or $\eta 7$ series. The $\omega 7$ series is therefore $CH_3(CH_2)_5CH=CH-\ldots$. According to this nomenclature, the linoleic acid, which is $CH_3(CH_2)_4CH=CH-CH_2-CH=CH-(CH_2)_7COOH$, or 9,12-octadecadienoic acid, can be described as $18:2\omega 6$. Since most of the natural fatty acids have *cis* double bonds, only the *trans* exceptions are labeled, such as the $18:2\omega 6$ *trans, trans* (the repetition of *trans* indicates that both double bonds are *trans*). Animal cells are unable to synthesize the $\omega 3$ and $\omega 6$ series (but they can make the $\omega 7$ and $\omega 9$ series) because they lack enzymes that can catalyze the desaturation toward the methyl end of fatty acids. Microorganisms and plant cells, on the other hand, are able to desaturate fatty acids toward the methyl end. The animal cells can only elongate and desaturate such fatty acids toward the carboxyl end. Therefore, the $18:2\omega 6$ and $18:3\omega 3$ that cannot be synthesized by animal cells are called *essential* fatty acids (EFS's). Table 4-2 shows the fatty acid composition of common food fats as found by the U.S. Department of Agriculture (USDA).

Glycerides

Glycerides are fatty acid esters of glycerol. The mono-, di-, and triglycerides are the most abundant. Their nomenclature is based on the labeling of the carbons 1, 2, and 3 (or α, β, α'):

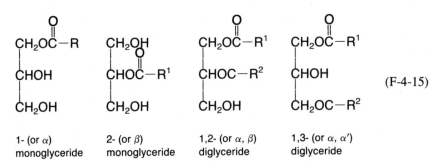

(F-4-15)

1- (or α)   2- (or β)   1,2- (or α, β)   1,3- (or α, α')
monoglyceride   monoglyceride   diglyceride   diglyceride

Oleic and palmitic are two important fatty acids:

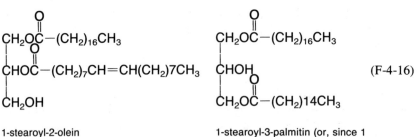

(F-4-16)

1-stearoyl-2-olein   1-stearoyl-3-palmitin (or, since 1 and 3 are equivalent, 1-palmitoyl-3-stearin)

TABLE 4-2 Fatty Acid Composition of Common Food Fats* (in 100 gm, edible portion)

Trivial Name	Fatty† Acid	Beef Tallow	Butter	Cocoabutter	Coconut Oil	Corn Oil	Cottonseed Oil	Olive Oil	Palm Oil	Peanut Oil	Soy Bean Oil	Lard (Pork)
	4:0		2.6									
	6:0		1.6		0.6							
	8:0		0.9		7.5							
	10:0		2.0		6.0							0.1
Lauric	12:0	0.9	2.3		44.6				0.1			0.2
Myristic	14:0	3.7	8.2	0.1	16.8		0.8		1.0	0.1	0.1	1.3
Palmitic	16:0	24.9	21.3	25.4	8.2	10.9	22.7	11.0	43.5	9.5	10.3	23.8
Stearic	18:0	18.9	9.8	33.2	2.8	1.8	2.3	2.2	4.3	2.2	3.8	13.5
	16:1	4.2	1.8	0.2			0.8	0.8	0.3	0.1	0.2	2.7
Oleic	18:1	36.0	20.4	32.6	5.8	24.2	17.0	72.5	36.6	44.8	22.8	41.2
	20:1	0.3						0.3	0.1	1.3	0.2	1.0
	22:1											
Linoleic	18:2	3.1	1.8	2.8	1.8	58.0	51.5	7.9	9.1	32.0	51.0	10.2
Linolenic	18:3	0.6	1.2	0.1		0.7	0.2	0.6	0.2		6.8	1.0

Source: Agriculture Handbook #8-4. 1984. U.S. Department of Agriculture. U.S. Government Printing Office, Washington, D.C.

* Average values.

† Number of carbon atoms: number of double bonds.

Some natural oils, such as fish oils, are more complex, since they contain glyceryl ethers. Waxes, which are esters of a long-chain alcohol and a fatty acid, such as

$$CH_3(CH_2)_{16}O\overset{\overset{\displaystyle O}{\|}}{C}-(CH_2)_{18}CH_3$$

also occur naturally. Beeswax and wall wax are both mixtures of such esters.

Ruminant milk fat is high in C_4 to C_{10} fatty acids (between 20% and 30% on a molar basis), while human milk is lower in C_4 to C_{10} acids and is higher in lauric (C_{12}) and myristic (C_{14}). Butter fat has nearly a hundred fatty acids, some with branched chains, as well as 2-, 3-, or 4-keto fatty acids, which contribute to the flavor and color of butter. Vegetable oils are quite different depending on the plant source and part; for example, erucic acid is not present in rape leaf oil, but is present in rapeseed oil (40% to 50%). This is important since the human heart lacks an enzyme for oxidizing the erucic fatty acid; thus, the accumulation of excess erucic (or other fatty acid) in the heart muscle membranes can cause heart muscle disfunction and heart disease. Plants in general produce linoleic and linolenic, whereas animal cells are unable to synthesize them. On the other hand, microorganisms are able to produce most kinds of fatty acids.

In the edible fat and oil industry, processing requires the availability of fats with a wide range of melting-points. These fats also need to have various plastic properties for diverse applications, such as chocolate (which is hard and brittle below $\sim 20°C$ and melts at $\sim 30°C$), salad and cooking oils (which are liquid even at $0°C$), or margarine (which can be spread immediately after being removed from the refrigerator).

Complex Lipids

Phosphoglycerides
Phosphoglycerides are a very important group of lipids that upon hydrolysis yield inorganic phosphorus and glycerol in addition to derived lipids.

This group includes phosphatidic acids, phosphatidycholine (PC), ethanolamines, serines, and inositols. Other phosphoglycerides are lysophosphatides, plasmalogens (monovinyl ether monofatty acyl phosphoglycerides), phosphonolipids, sphingolipids (with no glycerol backbone, but still in the phospholipid group), cerebrosides and sulfatides (including covalently linked carbohydrate groups), gangliosides, and cardiolipin (a polyphosphatidyl glycerol). Some of their structures are shown in Figure 4-23a and b.

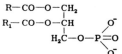

Phosphatidic acid

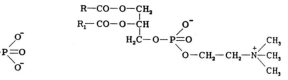

Phosphatidyl choline

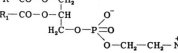

Phosphatidyl ethanolamine

Phosphatidyl serine

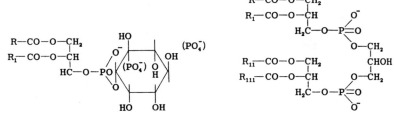

Phosphatidyl inositol

Cardiolipin

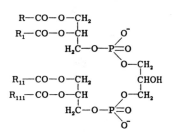

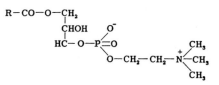

Lysophosphatidyl choline

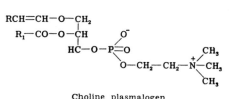

Choline plasmalogen

FIGURE 4-23a.   Chemical formulas of several complex lipids.

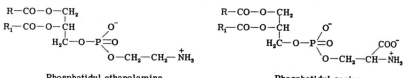

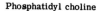

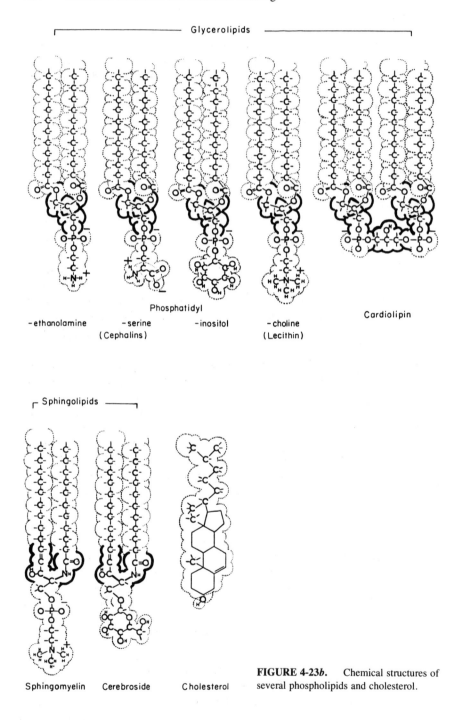

FIGURE 4-23b.   Chemical structures of several phospholipids and cholesterol.

Sterols

All sterols are based on a ring system, the *cyclopentanoperhydrophenantrene* (which contains a phenantrene-like ring as a subunit):

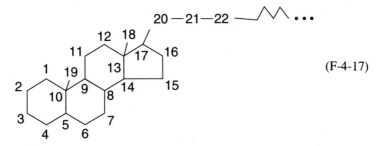

(F-4-17)

In this group are cholesterol, phytosterols (plant sterols), and terpenoids (for example, ursolic acid). Cholesterol has the chemical structure

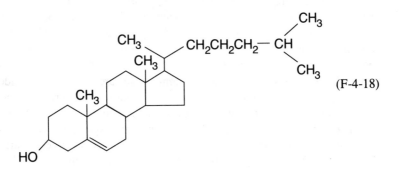

(F-4-18)

and occurs in most animal cell membranes, but it is not present in bacterial membranes. It can occur either as the "free" sterol, or with the OH group esterified with a fatty acid, in which case it is a *cholesteryl ester.* Blood serum, brain tissue, red blood cells, milk, and egg yolk contain significant amounts of cholesterol. In vivo, cholesterol is involved in fatty acid transport, in controlling the fluidity of cell membranes (higher cholesterol causes the membranes to be *less* fluid), as well as in the metabolism of adrenal corticosteroids and sex hormones. Cholesterol forms a characteristic green-colored complex upon reaction with acetic anhydride in $CHCl_3$ (the Libermann–Burchard reaction).

AMINO ACIDS

There are 21 different amino acids and two imino acids commonly found in natural proteins. The general chemical structure of amino acids is

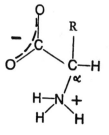

(F-4-19)

in their *dipolar ion* or *zwitterion* form; partially ionized forms are also present under very acidic or very basic pH conditions. The R-group, or side chain, gives an amino acid its identity (Fig. 4-24a and b, on pages 118, 119, and 120). The carbon atoms of the side chain are labeled sequentially by β, γ, δ, ϵ, η, ζ. The amino group attached to the α-carbon is called the α-*amino group*. The four bonds around the α-carbon are pointing at the vertices of a tetrahedron, as illustrated in Figure 4-25 (page 121). Because four different groups are surrounding the α-carbon, with the exception of R being H (glycine), the molecular geometry is asymmetric and there are two isomers possible, called *enantiomers*, an L- and a D-enantiomer. The L-configuration of the amino acids is the only one that occurs naturally in proteins.

The different properties of various amino acids are caused by the side-chain structure, R. There are four major types of R-groups: nonpolar, polar (but without a net charge), negatively charged, and positively charged. Note in Figure 4-24 that there are only two negatively charged R-groups, the carboxyls of glutamic (Glu) and aspartic (Asp) acids, and only three positively charged side chains, which belong to lysine (Lys), arginine (Arg), and histidine (His), while the less common hydroxylysine has also a positively charged R amino group. The imino acids, proline (Pro) and hydroxyproline (HPro) have the $-NH_2^+$ group linked to the α-carbon (instead of having the NH_3^+ group of the amino acids, so linked), and their rather special geometry has important consequences for the structure of the proteins in which they occur, as explained in the following section. Structures of the amino acids at pH 7.0, their names, and one- and three-letter abbreviations are shown in Figure 4-24a and b. The amino acids are classified by their side-chain polarity. Polar side chains are further classified as *neutral*, *basic*, or *acidic*.

The ionization states of amino acids are shown in the following structures. Structure (A) represents the undissociated (un-ionized) form of amino acids predominant at low pH (e.g., 1.0), whereas structure (B) is a dipolar ion (or "zwitterion"), present near neutral pH. At high pH, even the $-NH_3^+$ group releases a proton, becoming an $-NH_2$ group. Structure (C) shows the predominant form at high pH (e.g., 11). The conversion among these three forms (A, B, and C) occurs by the addition of H^+ of OH^-.

(A)

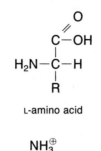

L-amino acid

(B)

$$\underset{\underset{\displaystyle R-CH-COO^{\ominus}}{|}}{NH_3^{\oplus}}$$

(C)

$$R-CH-C\begin{smallmatrix}O\\//\\\backslash\backslash\\O\end{smallmatrix}\ \ominus$$
with NH_2 above

PROTEINS

Proteins are high molecular weight heteropolymers of biological origin. They are made up of L-α-amino acids linked together through a *peptide bond* between the α-carboxyl group of one amino acid and the α-amino group of the adjacent amino acid. The schematic condensation reaction (R-4-4) is

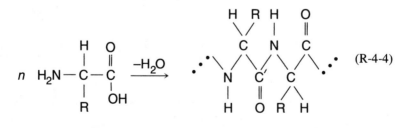

(R-4-4)

Simple proteins are proteins that upon hydrolysis, yield only amino acids and ammonia, while *conjugated proteins* are proteins that upon hydrolysis, yield other substances as well as those obtained from simple proteins. For example,

Lipoproteins contain lipids
Glycoproteins contain carbohydrates

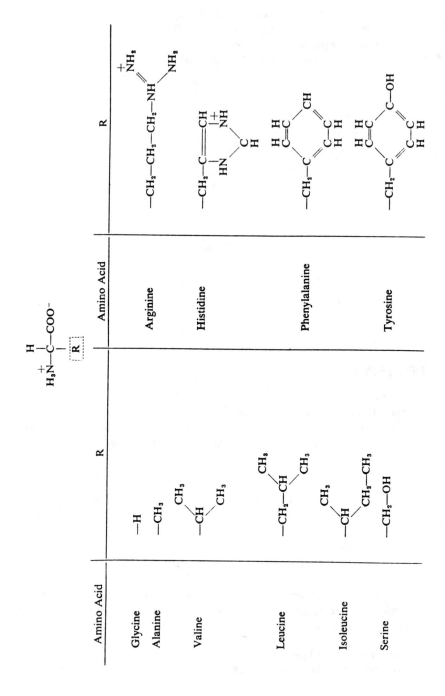

Amino Acid	R
Glycine	—H
Alanine	—CH₃
Valine	—CH(CH₃)CH₃
Leucine	—CH₂—CH(CH₃)CH₃
Isoleucine	—CH(CH₃)CH₂—CH₃
Serine	—CH₂—OH

Amino Acid	R
Arginine	—CH₂—CH₂—CH₂—NH—C(=NH₂⁺)NH₂
Histidine	—CH₂—C=CH—NH⁺=CH—HN—CH(H)
Phenylalanine	—CH₂—C₆H₅
Tyrosine	—CH₂—C₆H₄—OH

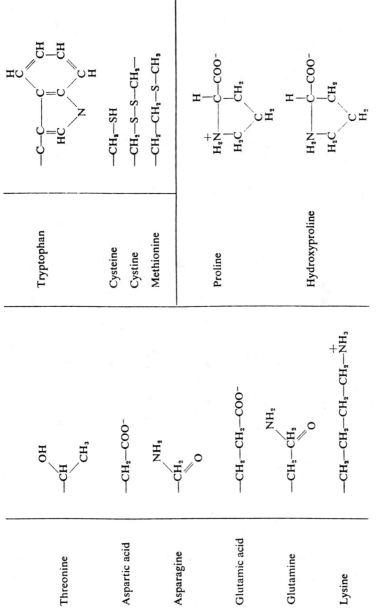

FIGURE 4-24a. Chemical formulas of amino acid residues.

119

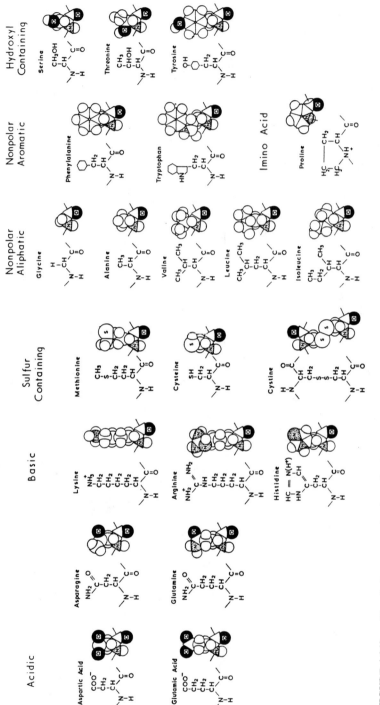

FIGURE 4-24b. Chemical structures of amino acids from natural sources. Also included is the structure of the imino acid proline and hydroxyproline. (From Finean, 1967, with permission.)

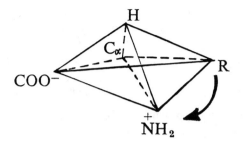

FIGURE 4-25.   Tetrahedral arrangement of hydrogen and three of the chemical groups surrounding the α-carbon.

Phosphoproteins contain phosphates
Metalloproteins contain metals such as iron and copper
Nucleoproteins contain nucleic acids

Locations of Various Amino Acid Residues in Protein Organization

The properties of R groups in conjunction with the amino acid sequence will largely determine their locations. These groups generally follow the rule: "**nonpolar in, polar out.**" Nonpolar R groups can, however, be exposed to the surface, depending upon the amino acid sequence of proteins and their environments. Neutral polar R groups are usually exposed to an aqueous medium, but can stay in the interior of proteins, especially when their functional groups are involved in hydrogen bonding. Charged R groups are almost always exposed to an aqueous medium because of their strong hydration.

Importance of Proteins

Proteins play crucial roles in fundamental life processes, and are present in the form of

1. Storage proteins (e.g., ferritin, casein, ovalbumin, lipovitellin, gliadins, glutenins, zein)
2. Contractile proteins (e.g., myosin, actin)
3. Structural proteins (e.g., collagen, elastin)
4. Protective proteins (e.g., antibodies, fibrinogen)
5. Hormones (e.g., insulin, growth hormones, adrenocorticotropic hormone)
6. Receptors (e.g., rhodopsin)
7. Enzymes
8. Carrier proteins (e.g., hemoglobin, myoglobin, transferrin, ceruloplasmin, plasma lipoproteins, plasma albumin)

The proteins important in the food industry or as food constituents are mostly storage proteins, enzymes, milk, and meat proteins. Some of the proteins listed later in the Basic Protein Structures section are also present as food constituents, or represent major sources of food proteins. A sampling follows:

1. Storage proteins are one of the most important sources of dietary proteins:

   Caseins account for approximately 80% of the total skim milk proteins
   Ovalbumin is an egg white protein
   Lipovitellin and lipovitellenin are egg yolk lipoproteins
   Gliadins and glutenins are major wheat proteins
   Conglycinin and glycinin are the two major soy storage proteins

2. Myosin and actin, which are involved in muscle contraction, are two major proteins of meat.
3. Collagen, which is a major structural protein of animals, is converted to gelatin by prolonged boiling.
4. Enzymes are widely used in food processing. Some enzymes, however, are responsible for food deterioration during storage.
5. Myoglobin, a carrier protein, is responsible for meat color, which is a major quality factor of meat and processed meat products.

Use of Requirements for Dietary Proteins

Dietary proteins supply nitrogen and amino acids, which the body uses to synthesize proteins, nucleic acids, the heme group for hemoglobin, as well as many other nitrogen containing compounds. For protein synthesis to occur *all* necessary amino acids must be present at the same time.

The *essential amino acids* are those that the body cannot make from other sources of nitrogen. They are histidine, isoleucine, leucine, lysine, methionine, phenylalanine, threonine, tryptophan, and valine.

The requirements for protein depend on the age, health, and sex of the person. Rapidly growing infants and pregnant, or lactating, women have higher protein requirements than do other groups of people. In underdeveloped areas of the world serious physical and mental retardation results from the lack of sufficient high-quality protein for young children. In general, the requirements are as listed in Table 4-3.

Quality of Food Proteins

Protein quality must also be considered among the requirements for dietary proteins. Animal protein is of high quality and contains all the necessary amino acids in the correct proportions for use by the body. Plant proteins (Tables 4-4

TABLE 4-3   Recommended Daily Allowances for Dietary Proteins

Group	Age	g/kg of body weight
Infants	0–6 months	2.2
	6–12 months	2.0
Children	1–3 years	1.3
	4–6 years	1.5
	7–10 years	1.2
Adolescents	11–14 years	1.0
	15–18 years	0.9
Adults	19+ years	0.8

and 4-5), on the other hand, generally lack sufficient quantities of certain essential amino acids, notably lysine, tryptophan, and methionine (Table 4-6). The cereals are the most widely used plant foods in the world; however, their protein content is low. Cereals are most often used in the form of flour, which usually consists of the endosperm fraction of the kernel only, so the protein contents of the kernel fractions must be considered separately (Table 4-5).

Proteins differ in their nutritive values; digestibility, the amount of protein in the food, and the amount of essential amino acids in the protein determine the quality of a food as a protein source (Table 4-6).

The limiting amino acids in the two most widely used cereals—wheat and corn—are lysine and tryptophan. Efforts to improve the amounts of these amino acids in corn have produced a variety, Opaque-2, that has much higher levels of both of these amino acids. The protein quality of this variety is close to 90%

TABLE 4-4   Approximate Protein Contents of Cereal Grains*

Type of Grain	Percent Protein
Wheat	
Cummon (hard)	12–13
Club (soft)	7.5–10
Durum (very hard)	13.5–15
Barley	12–13
Rye	11–12
Oats	10–12
Corn (dent)	9–10
Rice	7–9

*Moisture contents approximately 12%.

TABLE 4-5   Approximate Protein Contents of the Major Anatomical Parts of Wheat and Corn*

Grain	Kernel Fractions		
	Germ	Bran	Endosperm
Wheat			
Weight (percentage of kernel)	3	12	85
Protein (percentage of kernel fraction)	26	13	13
Corn			
Weight (percentage of kernel)	12	6	82
Protein (percentage of kernel fraction)	18	7	10

*Expressed on dry weight basis.

that of milk. Since plant proteins vary in amino acid content, it is possible to gain sufficient amounts of the essential amino acids by supplementing a cereal with legumes.

Plant seeds are a rich source of protein that is of a higher quality than that in the other edible parts of the plants (Table 4-7). In particular, the soybean seed is finding many food and nonfood uses in the United States (Table 4-8),

TABLE 4-6   Essential Amino Acids in Foods mg/100 gm Food

Food	Val	Leu	Ile	Thr	Met	Lys	Phe	Trp
Beef	886	852	1435	812	478	1573	778	70
Pork	616	897	608	583	321	961	496	162
Lamb	790	1203	778	583	383	1275	625	198
Chicken	1018	1472	1069	794	502	1590	800	205
Shellfish	765	1388	745	730	466	1262	645	184
Fish	1150	1445	900	861	539	1713	737	211
Whole milk	255	430	219	153	86	248	239	50
Eggs	847	1091	778	634	416	863	709	184
Wheat flour	493	840	435	321	174	248	581	128
Corn	461	1190	350	342	182	254	464	67
Oats	711	1012	526	462	234	517	698	176
Rye	561	728	414	395	172	401	522	87
Brown rice	433	648	300	307	183	299	406	98
Polished rice	408	581	296	234	150	255	342	95
Soybean	1995	3232	1889	1603	525	2653	2055	532
Peanut	1244	1876	990	764	267	1280	1125	155
Dry peas	1058	1530	961	914	205	1962	1033	202
Potatoes	93	121	76	75	26	96	80	33
Lettuce	71	83	50	54	24	50	67	10
Apple	15	23	13	14	3	22	10	3

TABLE 4-7   Approximate Protein
             Contents of Certain Seeds*

Species	Percent Protein
Legumes	
Soybean	32–46
Peanut	21–36
Peas, dried	20–25
Bean (*Phaseolus* spp.)	19–25
Oilseeds	
Sunflower seed	25–27
Sesame seed	24–26
Cottonseed	17–22

*Moisture contents of 3–10%.

though it has been used for food in the Orient for hundreds of years. This seed is important as a supplement to cereal products, not only for the amino acids it contains but also for the processing characteristics it gives the doughs.

Basic Protein Structures

The Peptide Bond

Formation of a protein chain occurs through peptide bonding and the elimination of one water molecule for each peptide bond formed (as shown in (R-4-4)).

TABLE 4-8   Uses of Soybean Protein in the
             United States, 1970*

Use Category	Million Pounds
Flours and grits	
Baked goods	60
Meat products	38
Beverages	12
Dry cereals and infant foods	9
Brewers' flakes	1
Pasta and macaroni products	1
Miscellaneous	7–12
Export	12
Corn-soy-milk blends (CSM)	90
Total	232–237
Concentrates	20–35
Isolates	25–40

*Current levels are substantially higher.

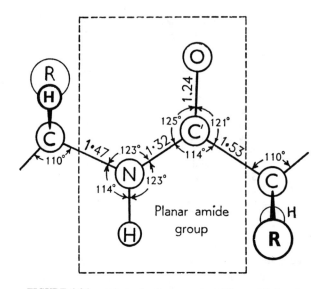

FIGURE 4-26.   The molecular geometry of the peptide bond.

(Note that C′ denotes the carbon atom that participates in the peptide bond and that C_α carbons are *not* involved in this bond (Fig. 4-26) but carry the amino acid residues, R). The peptide bond affects the electron distribution of the carbonyl and the nitrogen: the same electrons are redistributed over both the newly formed C′—N bond and over the carbonyl. The result is that both C′≡N and C′≡O acquire partial double-bond character (≡), and the carbonyl no longer has a "pure" double-bond character (C=), as represented classically in the amino acid monomer. Such a redistribution of bonding electrons has the very important result illustrated in Figure 4-27*a* and *b*.

The *sequence* of amino acids linked by peptide bonding into a protein chain is called *the primary structure* of that protein. Apart from the peptide bond, the only covalent bonding permitted within the same protein chain or between different protein chains is through *disulfide bonding*, which involves two —SH residues. There are, however, other noncovalent interactions between amino acid residues that cause the linear protein chain to fold. Such interactions are

1. Hydrogen bonding (for example, —C≡O · · · HN≡)
2. Hydrophobic interactions between nonpolar residues

Covalent (disulfide) bonding and the noncovalent interactions cause the occurrence of special arrangements, or structures, of the polypeptide chains.

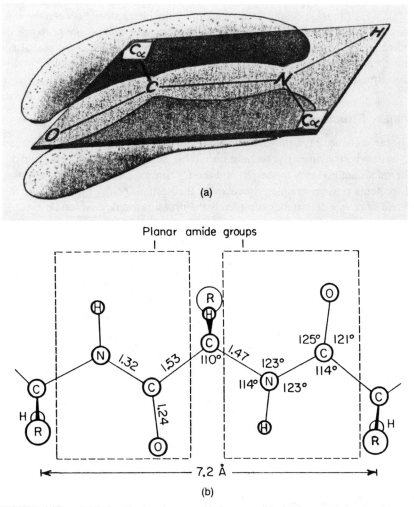

(a)

Planar amide groups

7.2 Å

(b)

FIGURE 4-27.   (a) Delocalized π-electron orbitals responsible for the peptide bond and its planar structure involving the C, O, N, and H atoms (from Zubay, 1983, with permission). (b) Representation of two amide group planes with rotation allowed around only the α-carbon.

Definition of Secondary, Tertiary, and Quaternary Protein Structures

Below we define the secondary, tertiary, and quaternary structures of proteins.

The *secondary structure* is a regularly recurring spatial arrangement of polypeptide chains as determined primarily by the hydrogen bonding between peptide bond elements. The *tertiary structure* is defined as the overall arrangement, or folding, of peptide chains as determined by the interactions between the side chains of the amino acid residues, by disulfide bonds, and by the presence of proline and hydroxyproline. Finally, the *quaternary structure* is the spatial arrangement of individual polypeptide chains in proteins with subunit or multichain structures (e.g., *arrangement of subunits* relative to each other).

Primary Structure of Proteins

A linear sequence of amino acids in proteins linked by peptide bonds is called the *primary structure*. By convention, linear sequences are shown with the N-terminal amino acids to the left and the C-terminal amino acids to the right. All proteins possess unique sequences of their amino acids. Furthermore, the *protein sequence* determines completely the *protein structure* (secondary, etc.).

Three-Dimensional Structure of Proteins

A protein, such as any food protein (caseins or soy proteins, for example), may have within itself both regular or disordered structures, as well as almost random regions. Such regions within a protein are called *structural domains*. Examples of well-ordered or regular structures occurring in proteins are shown in Figure 4-28a to f. Because of the highly ordered arrangement of the amino acid residues such proteins are fibrous, that is, they form fibers.

β-bends (β-turns) have the following characteristics. They occur between chains that have an ordered structure (i.e., helix and sheet), but are *not* a part of the ordered structure, and they can determine the direction of two such ordered chains, and provide a 180° reversal of the chain direction. A β-turn is usually located at the surface of the native globular proteins and has one intrachain hydrogen bond. Four amino acid residues make one β-bend (e.g., Fig. 4-29).

Regular Protein Structures: Important Examples

Myoglobin

Myoglobin, whose role is to store oxygen in muscle, is the first protein structure determined by X-ray crystallography (Fig. 4-30a). It is an excellent example of a highly ordered α-helical structure, and its content of α-helix is about 85%, as shown in Figure 4-30b. (The effect on the computed structure of increasing resolutions is shown in Fig. 4-30c to f.) Note that there are eight helical regions in myoglobin, which form a "pocket" (Fig. 4-31) that surrounds the heme (Figs. 4-30h and 4-31). It is the iron ion, Fe^{2+}, in its ferrous form that binds the oxygen molecule (Fig. 4-31). The ferrous ion is octahedrally coordinated

(*Text continues on page 135.*)

FIGURE 4-28. Regular structures of polypeptide chains and proteins.

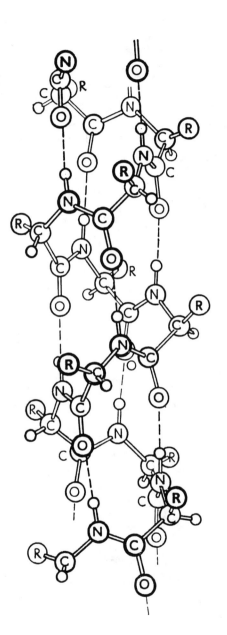

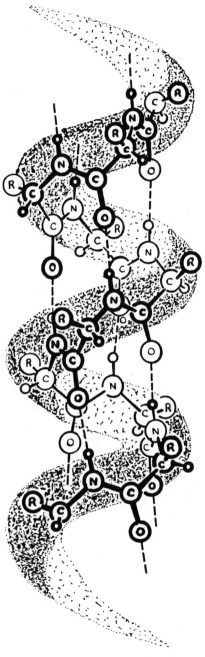

FIGURE 4-28a. Ball-and-stick model of an α-helical structure.

FIGURE 4-28b. Pauling and Corey model of α-helix.

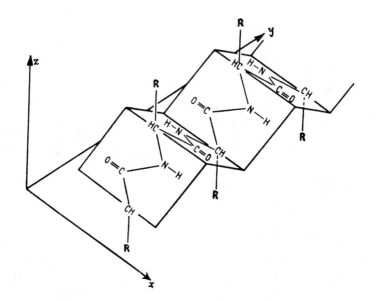

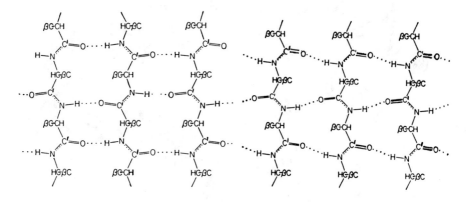

FIGURE 4-28c.   Schematic representation of a β-pleated sheet structure.

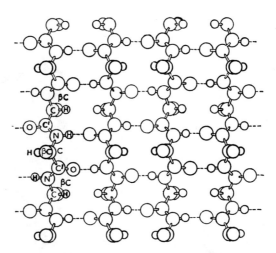

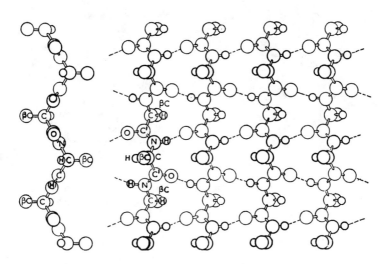

FIGURE 4-28d.   Ball-and-stick models of parallel (*top*) and antiparallel (*bottom*) β-pleated sheets.

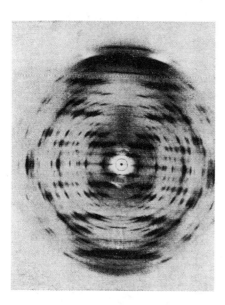

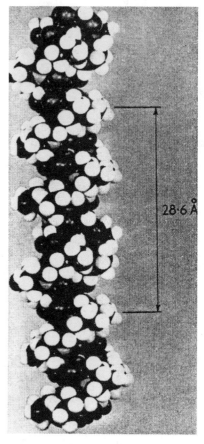

28·6 Å

FIGURE 4-28e.   X-ray diffraction pattern of oriented collagen fibers.

FIGURE 4-28f.   CPK model of the triple helix of collagen corresponding to the X-ray diffraction pattern in part (e).

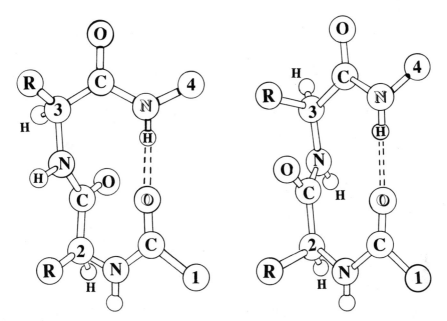

FIGURE 4-29.   Beta-turn structures, with glycine occurring often at position 3 in the turn because of the small size of its residue.

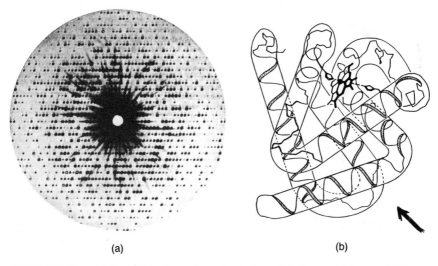

(a)   (b)

FIGURE 4-30.   (a) X-ray diffraction pattern of a single crystal of sperm whale myoglobin. (b) Schematic model of the three-dimensional conformation of myoglobin, showing the α-helical folding of the protein chain, the tertiary structure, and the heme group.

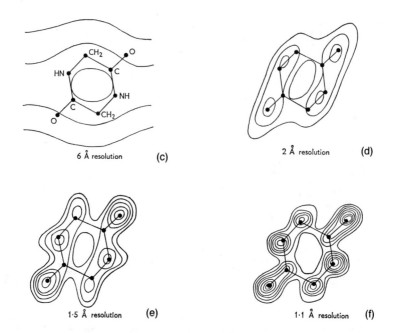

FIGURE 4-30 (continued).   (c) Reconstructed electron density of a side chain in myoglobin. (d) to (f) The structure shown at increasing resolutions (from Wilson, 1966, with permission).

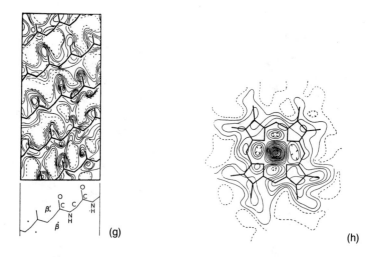

FIGURE 4-30 (continued).   (g) Electron density along the polypeptide chains reconstructed from the X-ray diffraction pattern in part (a). (h) The effect of resolution on the electron-density reconstruction for the heme group (from Wilson, 1966, with permission).

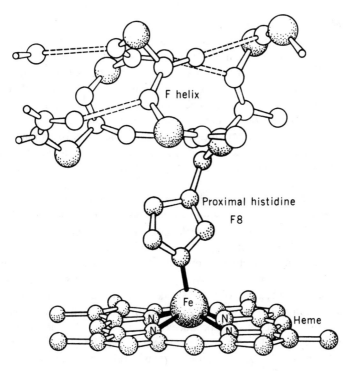

FIGURE 4-31.   Coordination of the iron in the heme group and its bond to the proximal histidine F8 in myoglobin.

(that is, its coordination number is six), as shown above in Figure 4-31. The polypeptide-chain electron density derived from the X-ray diffraction pattern in Figure 4-30a is shown in Figure 4-30g, and a side chain group electron density at different resolutions is shown in Figure 4-30c–f.

β (Beta) -Pleated Sheets

A second variety of fibrous proteins are the *silk fibroins*, which are produced by insects. Silks are structurally composed of aligned and stacked antiparallel β sheets, in which the side chains point alternately upward and downward from the plane of the sheet, all the glycine residues are arranged on one surface of each sheet and all the substituted amino acids are located on the other side of the sheet. Two or more such sheets can therefore be packed closely together to form an arrangement of stacked sheets in which two adjacent glycine-substituted or alanine-substituted sheet surfaces interlock with each other (Fig. 4-28a and b). Because of the extended conformations of the polypeptide chains in the β-sheets and the interlocking of the side chains between sheets, silk is a mechanically rigid material that tends to resist stretching.

Triple-Helix

A special type of structural protein is *collagen*. This is a particularly rigid and inextensible material that is a major constituent of tendons and many connective tissues. Analysis of collagen amino acid sequences shows them to be characterized by a repetitious tripeptide sequence, Gly-*X*-proline or Gly-*X*-hydroxyproline, where *X* can be any amino acid and hydroxyproline is a hydroxylated derivative of proline. Owing to the repeating proline residue, collagen polypeptide chains cannot adopt either an α-helical or a β-sheet conformation. Instead, individual collagen polypeptide chains tend to assume a left-handed helical conformation in which successive side-chain groups point toward the corners of an equilateral triangle when viewed down the polypeptide chain axis ('end-on') (Fig. 4-32*a*). The glycine occurrence every three residues is strictly required because there is no space for any other amino acid residue inside the triple helix, where the glycine *R* groups (H) are located. Each collagen helix is already extended (Fig. 4-32*b*), so it cannot easily stretch further like the α-helix. In contrast to the α-helix, formation of a single-chain collagen helix *cannot* be accompanied by the formation of hydrogen bonds among residues *within* each polypeptide chain (see, e.g., Fig. 4-28*e* and *f*). Instead, three collagen helical chains associate in a three-stranded "cable" with hydrogen bonds *between* each chain and its two neighbors. This produces a highly interlocked fibrous structure that is especially suited to its biological role of providing rigid connections between muscles and bones, as well as structural reinforcement in the skin and connective tissue.

Although additional types of protein exist, attention has been here focused on three arrangements whose structural properties are currently best understood. Two of these, the α-keratins and the silks, incorporate polypeptide secondary structures that also occur often in globular proteins. In contrast, collagen is a protein that evolution has developed to play a special, structural role.

Ramachandran Plots

The three-dimensional structures of proteins, and especially the *regular secondary structures*, are readily specified by employing two dihedral angles, ϕ and ψ. The ψ angle defines the rotation of the carbonyl group around the $C^{\alpha}-C'$ bond, whereas the ϕ angle defines the rotation of the NH group around the $C^{\alpha}-N$ bond. A plot of ψ against ϕ is called a *Ramachandran plot*, two examples of which are given in Figure 4-33*a* and *b*. The Ramachandran maps graphically define the regions of allowed secondary structures. The points in Figure 4-33*b* represent specific values of the ϕ and ψ angles for which the potential energy of the corresponding conformation is *minimal*, and, therefore, they represent *stable* conformations.

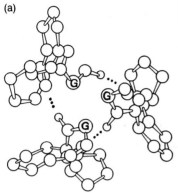

(a)

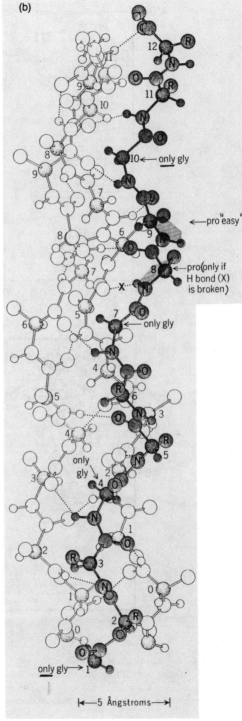

(b)

12
11
10 ← only gly
← pro "easy"
9
8
← pro (only if H bond (X) is broken)
7 ← only gly
6
5
only gly → 4
3
2
1
0
only gly → 1

|←—5 Ångstroms—→|

FIGURE 4-32. The basic coiled-coil structure of collagen. Three left-handed single-chain helices wrap around one another with a right-handed twist. (*a*) View from the top of the helix axis. (*b*) Ball-and-stick single-collagen chain. Note that glycines are all on the inside (with permission from Zubay, 1983).

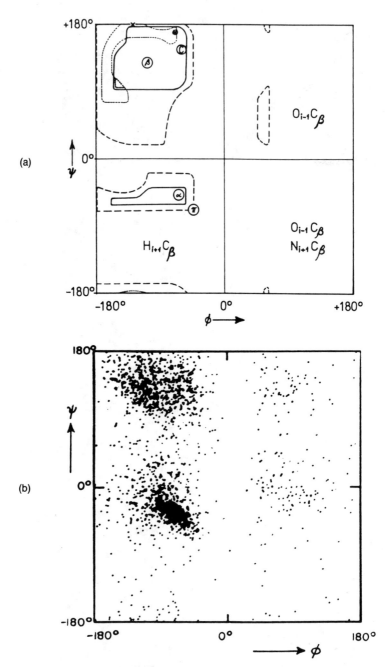

FIGURE 4-33. (*a*) Computed Ramachandran plot for polyalanine. (*b*) Experimental Ramachandran map derived from known crystal structures of proteins obtained from X-ray diffraction studies. Each point in this plot represents a couple (ψ_0, ϕ_0) of Ramachandran angles for specific residues in such protein crystal structures (from Schulz and Schirmer, 1978, with permission).

Physiochemical Properties of Proteins

Ionic Properties of Proteins

Acid-Base Properties of Proteins
Neither the α-amino nor the α-carbonyl groups in peptide linkages can ionize; only the α-amino group at the N-terminal and α-carboxyl group at the C-terminal can retain their charges. Their pK' values are shifted approximately one unit toward neutrality.

The pK' values of the ionizable side chains are similar to those of the corresponding amino acids, unless they are subject to electrostatic interactions or are present in hydrophobic environments.

Isoelectric and Isoionic Points
The *isoelectric point* (pI) is defined as the pH of a protein solution at which the net charge on the protein is equal to zero; that is, the pH at which the protein will not migrate in an electric field. The isoelectric point is a function of the nature and concentration of the buffer, and in general, of any other solutes present in which the pI is measured. The pI of proteins is affected by the presence of ions and solutes that are capable of binding to the protein.

A protein solution that contains no ion other than those *intrinsic* to the protein itself and those derived from the dissociation of water is called an *isoionic solution*. Isoionic solutions can be prepared by passage of the protein solution through a column containing both anion- and cation-exchange resins (a mixed-bed resin) so that all ions other than H^+ and OH^- are removed. Isoionic solutions can also be prepared by extensive dialysis against distilled water. The pH of an isoionic protein solution is termed the *isoionic point* of that protein. Factors influencing ionic properties of proteins are salts, cations, and anions.

Food Proteins

Vegetable proteins include cereal and soy proteins/soy protein preparations. Several types of soy protein preparations that are made in large quantities (Table 4-8) by the U.S. food industry as food ingredients are listed below.

Preparation	Production (in tons)
Soy flours	320×10^6
Concentrates	36×10^6
Isolates	35×10^6

11S Soy Globulin
Glycinin, or 11S, soy globulin constitutes about 30% of the total soy protein and is located primarily in the protein storage bodies of the soybean. Soy stor-

age proteins do not seem to have a specific biological function, but are used as a source of nitrogen during germination. These are synthesized on the rough endoplasmic reticulum of cells at later stages of development, after which they are deposited in the protein bodies for storage. These bodies are spherical organelles about 5 μm in diameter that are bound by a membrane surrounding the protein.

The *molecular weights* of the glycinin subunits are, respectively,

$M_w \sim$ 37,000 for the acidic subunits (pI: 5.2, 5.4, and 4.8)
$M_w \sim$ 20,000 for the basic units (pI: 8.0, 8.3, and 8.5)

There are six pairs of acidic (A) and basic (B) subunits held together by *disulfide bonds*. It has been proposed that the secondary structure of glycinin is about 6% α-helical (very low!), 37% β-pleated sheet, and the rest is about 58% random coil. This, however, may not be the native secondary structure!

The *quaternary structure of glycinin* was derived from electron microscopy (EM) and small-angle X-ray scattering (SAXS) studies, and seems to involve two stacked hexamers of A and B subunits.

The amino acid (AA) composition of glycinin is shown in Table 4-9.

TABLE 4-9   Amino Acid Composition of Glycinin and Its Subunits

| Amino Acid | Glycinin | Percentage of Total Residues Acidic | | Basic |
		A2	A3	
Asx	11.8	12.4	12.0	14.0
Thr	4.2	3.6	4.1	4.7
Ser	6.6	4.8	7.1	8.6
Glx	18.8	25.4	25.0	13.4
Pro	6.3	6.3	8.9	5.3
Gly	7.8	8.8	7.8	6.7
Ala	6.7	5.3	2.9	7.5
Val	5.6	4.5	4.6	6.6
Met	1.0	1.7	0.6	1.2
Ile	4.6	4.5	3.2	4.8
Leu	7.2	5.9	5.7	9.0
Tyr	2.5	1.9	1.5	2.8
Phe	3.9	3.6	3.1	4.6
His	1.8	0.8	3.7	1.6
Lys	4.1	4.4	3.9	3.3
Arg	5.9	6.7	5.8	5.4
Cys	1.1	1.3	0.9	0.7
Trp	0.7	—	—	—

Sulfhydryls and Disulfide Bonds in Glycinin

Because the native glycinin is not readily hydrolyzed by proteases, it is thought that the native conformation is a tightly folded one, which is presumably maintained by disulfide bonds because denatured glycinin does *not* renature in the presence of mercaptoethanol, or DTT, but does renature about 70% if these disulfide bond-breakers are not present. Free SH groups of ~1.1 mole/mole protein are exposed to solvent and 10.1 SS groups are not exposed. The total, however, is about 37 S—S bonds.

Denaturation of Glycinin by Urea or Guanidine Hydrochloride

8M urea or 6M guanidine hydrochloride completely dissociates the 11S complex of glycinin and causes denaturation. After removing the 8M urea, renaturation occurs to about 70% of the basic structure.

7S/11S Gelling Behavior: Heat Effects

At low ionic strength glycinin dissociates into subunits at about 70°–80°C (heating for 10 min), while at high ionic strengths dissociation into subunits occurs between 90°C and 100°C.

Gels can be formed containing both glycinin and conglycinin, and the most important role played in holding these gels together is that of *disulfide bonding* (irreversible gelling). Hydrophobic interactions seem to be less important in this case, and hydrogen bonding may be involved in reversible gelling of these soy proteins.

Wheat and Corn Proteins

In this section we first discuss the distribution of protein in cereals and in the various parts of the grain. For example, at natural moisture levels, the protein content of all the common cereals averages around 10%, although individual samples of particular cereals might contain as little as 6% or more than 20% (e.g., Table 4-4).

The protein is distributed nonuniformly among the morphological tissues of the grain, the highest concentrations occurring in the outermost, or subaleurone, part of the so-called "starch endosperm" and in the germ and the aleurone layer of the endosperm. The inner endosperm has a lower protein content than that of the whole grain, and there is very little protein in the pericarp.

By manual dissection of whole grains and microanalysis of the parts, the distribution of protein in wheat and in corn was obtained (see Table 4-5).

Classes of Cereal Proteins

Proteins in cereal endosperm have been classified structurally in numerous ways. None of these classifications is entirely satisfactory, but all have made some contribution to our knowledge of cereal proteins.

Treatment of wheat flour successively with water, dilute salt solutions, 70% aqueous ethanol, and dilute acid or alkali removes protein fractions that have been called, respectively, *albumin, globulin, prolamin (gliadin)*, and *(glutenin)*, since Osborne first carried out this fractionation in 1907. (Note that the Osborne classification is the most widely used in spite of its proven shortcomings.) Since this early work, cereal protein chemists have shown that *none of these fractions is a pure protein*. The albumins and globulins, however, have lower molecular weights and faster electrophoretic mobilities on starch gel than the other fractions. The gliadins and glutenins constitute about 80–85% of the total endosperm protein, and together with small amounts of other proteins, lipids, and starch, make up the "gluten," or hydrated protein, that is obtained as a coherent mass when a flour–water dough is kneaded under running water to remove the starch and water-solubles. Although gliadin is soluble in aqueous alcohol, and glutenin in dilute acids and alkalis, these two fractions together are often referred to as *insoluble protein* (meaning insoluble in water and salt solutions). *Soluble* protein generally refers to albumin and globulin.

Wheat Gluten Proteins: Gliadins
Purified gliadins can be isolated (Baianu, 1981), for example, from single-variety *Flanders* flour by a simple four-step preparation procedure at 4°C.

1. Extraction of albumins and globulins with 1-M NaCl solution
2. Extraction of gliadins and certain glutenins with 50% propanol
3. Repeated dialysis of extracted gliadins and glutenins with distilled water
4. Centrifugation at 50,000 g for 2 h to eliminate the precipitate (mostly glutenins)

The clear supernatant obtained at stage (4) can be lyophilized and stored at $-20°C$.

Sodium dodecylsulphate–polyacrylamide gel electrophoresis (Baianu, 1981; Baianu et al., 1982) of the reduced supernatant indicates that the sample consists of a representative population of gliadins from *Flanders* flour, without any trace of glutenin or other protein subunits with molecular weights in excess of 75,000 Daltons (Fig. 4-34a). This is confirmed by additional observations using conventional electrophoresis on starch gels and by gel filtration chromatography (Fig. 4-34b). The latter technique also indicates a decrease in the relative amounts of ω-gliadins in the separated gliadins by comparison with the control.

Milk Proteins: Subfractions
and Their Electrophoretic Behaviors

Electrochemical Properties of Milk Proteins
Due to the presence of acidic and basic groups in the protein molecules, milk proteins are amphoteric in nature, that is, they can possess either a net positive

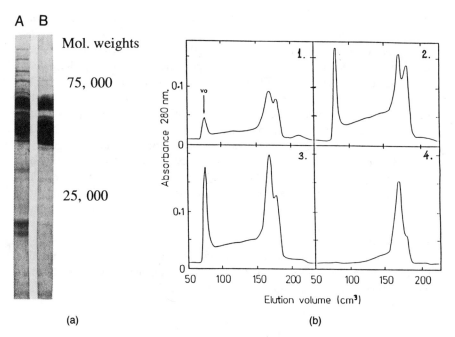

FIGURE 4-34.   (*a*) Sodium dodecylsulphate–polyacrylamide gel electrophoresis of wheat gliadin extracts from the single-variety *Flanders* of wheat grain; tracks: (A) 50% propanol:water extracts, dialized, and (B) Molecular weight, protein standards (Sigma, Co.) (Baianu, 1981). (*b*) Gel filtration chromatography of wheat proteins in *Maris Huntsman* variety: flour (1), gluten (2), glutenin fraction (3), and gliadin-enriched fraction (4) (following 2-M urea extraction) (Schofield and Baianu, 1982).

or negative charge in solution, depending on their environment. This charge depends on the extent of the dissociation of these groups, and the character and number of any bound ions. The common methods for investigating the electrochemical nature of milk proteins are their titration curves and electrophoretic behavior.

Electrophoretic Behavior
Electrophoresis has been extensively used for the identification and characterization of the various milk proteins. With the advent of zonal electrophoresis on polyacrylamide, or starch gels, with their higher resolving power, this technique has largely replaced free-boundary electrophoresis. The relative mobility of the proteins on these gels not only reflects the charge of the proteins but also their size and configuration in the particular medium employed. For example, a 6-M urea, 10% starch gel with a 0.76-M tris-citrate buffer, pH 8.6, separates the genetic variants of α_{si}-casein and β-casein, but because of the presence of disulfide polymers, κ-casein either remains in the slots or streaks. To phenotype the κ-caseins their gel system must be changed to 5 M-urea, 10% starch with

0.03-M mercaptoethanol, to reduce the disulfide bonds and form the "monomers" that can then be identified.

The physical state of the milk proteins in their natural environment has been investigated by their free-boundary electrophoretic behavior with the natural-protein-free-milk systems (NPFMS) as the buffer. At equilibrium, the NPFMS contains all of the dialyzable constituents of the milk *at the same activities* that they possess in the milk (see Chap. 6 for a definition of the concept of "activity"). A typical electrophoretic pattern of skim milk and its fractions in NPFMS is shown in Figure 4-35, together with the pattern of its ultracentrifuged fractions. These patterns indicate that most of the casein present exists in the form of micelles, but at 2°C some of the caseins exist in nonmicellular form. The mobilities of the proteins vary with the sample of milk and with its previous treatment.

Electrical Properties
The presence of ionizable groups causes the charge on proteins. The ionizable groups in α- and β-caseins, and β-lactoglobulin are phosphoserines, lysines, and arginines, as well as glutamic and aspartic acids.

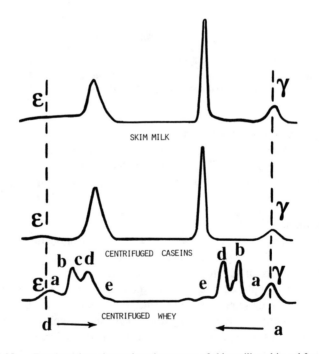

FIGURE 4-35.   Free-boundary electrophoretic patterns of skim milk and its subfractions obtained by ultracentrifugation (55,150 × g for 10 h at 1°C) with its natural-protein-free-milk system as a buffer (Whitney, 1977).

Major Properties of the α_{s1}-Casein
The molecule is relatively hydrophobic and the distribution of proline restricts the formation of a secondary structure. The α_{s1}-casein associates mainly by hydrophobic bonding. Therefore, α_{s1}-casein is sensitive to precipitation by calcium ions, probably due to the interaction of Ca^{2+} with the phosphorylated serine residues, and possibly with the ionized carboxyl groups.

In α_{s1}-casein the molecular weight is approximately 23,500 Daltons for 50–55% of the whole casein. These caseins have the greatest number of anionic phosphate groups. The α_{s1} casein has four genetic variants, A–D. A and D are very rare; B is the most common.

β-Caseins

In β-caseins the molecular weight is approximately 24,000 Daltons for 30–35% of the whole casein, although they have fewer phosphorylated serine residues than the α_{s1}-caseins. There are six variants, with A^2 being the most common. The hydrophobic nature of the associations of β-casein is shown by the effects of temperature, pH, and concentration. At a temperature $\leq 8\,°C$, or at high pH, β-casein dissociates into monomers; at higher temperature and pH near neutrality, β-casein associates into threadlike polymers. The β-casein is not as sensitive to calcium as the α_{s1}-casein, since instead of a precipitate, it forms a suspension.

The outstanding physical characteristic of β-casein is its temperature-dependent association. If one examines its primary structure, it is readily apparent that most of the negative charges exist in the 21 N-terminal amino acid residues at neutral pH values, and that the balance of the molecule is essentially neutral and contains a large number of hydrophobic residues. The molecule itself possesses a large hydrophobicity, and a large proportion of proline residues (~ 17%) that should result in an even greater number of nonpolar groups exposed at the surface than α_{s1}-casein. Therefore, the tendency for hydrophobic bonding should be great, and an increase in association with an increase in temperature is to be expected.

"γ"-Caseins

In "γ"-caseins the molecular weight is 11,500–20,000 Daltons for 5% of the whole casein. The "γ"-casein sequences represent segments of β-casein, however. They may possibly be derived from β-casein by a limited proteolysis by casein-associated endogenous proteases. Only "γ"-casein contains a single phosphorylated serine.

κ-Caseins

The κ-caseins make up that portion of the α-casein fraction that is soluble in 0.4-M $CaCl_2$ at pH 7.0 and 4°C. They are capable of stabilizing the α_s-caseins. When isolated, they appear to consist of a mixture of polymers of a monomeric unit of molecular weight of approximately 19,000 Daltons, and are held together by intermolecular disulfide bonds. However, free SH groups are detected in κ-casein when all of the calcium has been removed by ethylenediaminetettraacetic (EDTA) or oxalic acid, suggesting the possibility that the reduced monomers rather than the disulfide-linked polymers are closer to the native form of κ-casein. More work is needed before a definite conclusion can be reached on this matter.

β-Lactoglobulins

Since the time β-lactoglobulins were first isolated from milk, extensive studies have been made of their structure and configuration. Most early workers in the field obtained a molecular weight of 36,000–40,000 Daltons for β-lactoglobulin in the neutral pH range. However, they obtained a molecular weight of 17,000 Daltons from surface-film measurements in 20% $(NH_4)_2SO_4$ solutions, and therefore under these conditions, β-lactoglobulin dissociated into two fragments.

Green and Aschaffenburg (1959) determined the geometric structure of β-lactoglobulins A and B in the crystalline state (Fig. 4-36a–d), and found both proteins to consist of two spheres of radius 17.9 Å joined in such a way that the centers are 33.5 Å apart with a twofold axis of symmetry (see Fig. 4-36c).

Casein Micelles

The structure of the casein micelles in milk has long been of interest to investigators in the field of milk proteins. Considerable information has been accumulated with regard to their structure. From electron microscopic, sedimentation, and inelastic light-scattering investigations, the micelles were found to have molecular weights ranging from 1×10^7 to slightly greater than 6×10^8, with a weight average molecular weight of approximately 2.5×10^8 (Fig. 4-37). Their hydrodynamic radii range from approximately 300 to 1300 Å with a medium range value of 840 Å, and their sedimentation constants vary from approximately 100 to 1500 svedberg (S) units with a medium range value of 500 S.

The major properties of casein micelles are as follows. Approximately 80% of casein is present in the form of micelles, which contain 93% protein, 2.8% calcium, 2.3% organic phosphate, 2.9% inorganic phosphate, 0.4% citrate, and low levels of Mg, Na, and K. The ratio of the casein proteins is $\alpha_{s1} : \beta : \kappa = 3 : 2 : 1$ in most micelles; in smaller micelles the κ content is higher. Organic phosphate is attached to the Ser residues of α_{s1} of β-casein. They act to bind calcium ions. Inorganic phosphate, on the other hand, is in a complex with calcium and citrate as a colloidal moiety dispersed through the micellar phase. The micelles are 50–300 nm in diameter, and their molecular weights range from 1×10^7 to approximately 6×10^8 (average 2.5×10^8).

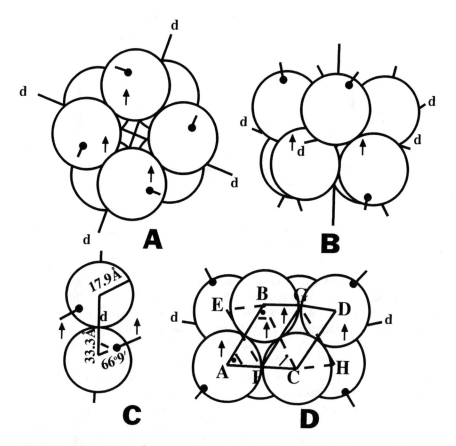

FIGURE 4-36.   Staggered structures of the octamer of β-lactoglobulin A; T = tetrad axis and d = dyad axis of symmetry (according to Green, 1964, and Timasheff and Townend, 1964, cited by Whitney, 1977). A to D are different views of such structures.

Evidence indicates the micelles are porous and highly solvated, but the exact structure is not yet known. Hydrophobic forces play a great role in micelle organization. Ion-pair formation between κ- and α_{s1}-caseins may contribute to the stability of the micelles. Calcium is necessary to micelle stability, since its removal results in breakdown of micelles into subunits. Colloidal calcium phosphate is also present in the micelles in association with caseins. Micelles consist of subunits 10–20 nm in diameter, containing 25–30 casein monomers.

The action of inter- and intramolecular ionic bonds is less certain. If we consider the polar groups in the sequences of the caseins, the possibility for ion-pair bonds in the apolar environments in the interior of the micelle is apparent. Carbamylation of five of the nine lysine residues of κ-casein destroyed

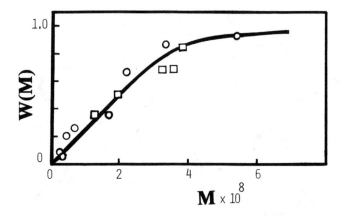

FIGURE 4-37.   Weight fraction of casein micelles with sizes (masses) determined by sedimentation and from diffusion measurements, less than or equal to the indicated value M.

its ability to stabilize the α_{s1}-casein; thus, in addition to the hydrophobic forces, ion pairs may be a factor in micelle stability. The association of *calcium* with the caseins must also be involved in micelle stability, though it is difficult to separate its effect from that of the colloidal calcium phosphate.

As indicated previously, the caseins in their "monomeric" form possess little secondary structure, but upon association conformational changes might occur. Undoubtedly some conformational changes take place upon micelle formation, and therefore must be involved in the stabilization of the micelle. Since numerous proteins are partially stabilized in their configurations by hydrogen bonds, as in the case of α-helices and β-pleated sheets, the possibility of hydrogen bonds between the components of the micelles cannot be eliminated. However, as yet, no direct evidence of their existence has been obtained.

The presence of *disulfide bonds* in the "polymer" forms of κ-casein undoubtedly carries over to the casein micelles, and therefore might contribute to their general structure. However, since the reduced and alkylated form of κ-casein also has the ability to stabilize α_{s1}-casein against precipitation by calcium, their contribution to the formation of the casein micelle may not be great.

Proposed Models for Casein Micelles

Coat-Core Model

In the core-coat model α_{s1}- and β-caseins are associated radially in rosettes to form the core (Fig. 4.38a). The κ-casein associates with the surface of the core, producing a hydrophilic coat (Fig. 4.38a). Micelle size is governed by the amount of κ-casein. However, this model does not allow equal access to all components by carboxypeptidases, in contrast to experimental results.

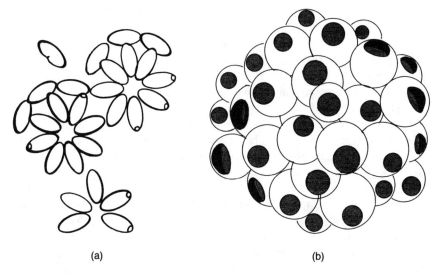

FIGURE 4-38.   (*a*) Core-coat model of a casein micelle. (*b*) Micelle asymmetric subunit model of casein micelles; the white spheres represent associated α_{s1}- and β-caseins, whereas the dark patches represent the hydrophilic κ-caseins facing the aqueous solution.

Uniform Subunits Model

In the uniform subunits model subunits consist of α_{s1}-β-casein polymers covered with layers of κ-α_{s1}-casein unit. The fact that smaller micelles seem to contain a larger proportion of κ-casein argues against the uniform subunit model.

Asymmetric Subunits Model

In the asymmetric subunits model, κ-casein is localized on one portion of the subunit surface, producing a hydrophilic patch (Fig. 4-38*b*). The remaining surface consists of α_{s1}- and β-casein, and can become involved in the association with similar surfaces on other subunits. This model provides for open channels throughout the micelle, producing *porous structures* accessible to monomeric caseins, rennin, and other proteins. Micelle growth is limited as the radius of curvature increases, since less hydrophobic surface is available for interaction with additional subunits. Smaller micelles would be produced when the subunits contain a greater amount of κ-casein. Since this would result in a larger hydrophilic patch on the subunits, the limit for micelle growth would be reached sooner than with the usual κ-casein content. The asymmetric subunit model, so far, provides the best explanation of the observed properties of the micelles. Considerable evidence exists that indicates that their structure is porous and highly solvated, with approximately 1.9 g of water per gram of protein. Carboxypeptidase A (mol. wt. 40,000 Daltons) is able to penetrate into the center of the casein micelle, since it quantitatively removes the carboxyl-terminal residues from all of the casein in the micelle.

While the composition of casein micelles is a function of temperature and the natural variability of milk composition, the micelles have been estimated to contain 50% α_{s1}-casein, 33% β-casein, and 15% κ-casein. In addition to proteins they are known to contain colloidal calcium phosphate with a calcium-to-phosphate ratio of 1.5–1.8.

To understand why the casein micelles appear to be stable in milk, one needs to examine the possible molecular interactions involved. We have seen earlier that the caseins are highly hydrophobic and quite capable of association through hydrophobic interactions, involving not only similar caseins but also different casein species.

Several investigators have noted that β-casein, and to a lesser extent κ- and α_{s1}-casein, diffuses out of the micelle at low temperature (1°C). This behavior is consistent with the hydrophobic character of these proteins, since the outstanding feature of β-casein is its temperature-dependent association. One-to-one association, which is largely hydrophobic, can occur between α_{s1}- and κ-casein. Dissociating agents, such as sodium dodecyl sulfate, guanidine-HCl, and urea, which overcome the hydrophobic association of the individual caseins, tend to disrupt the casein micelles, reducing the micelles to subunits approaching 100 Å. Thus, the stability of the casein micelle must be due at least in part to hydrophobic interactions.

Muscle Proteins

Skeletal muscle consists of long, cross-striated muscle fibers. The fiber comprises the myofibrils, the sarcoplasmic matrix, and some small structural elements including the sarcoplasmic reticulum, the mitochondria, and the nucleus (Fig. 4-39a and b). The (partially crystallized) protein gel of myofibrils, which contains the contractile substance, amounts to about 80% of the fiber volume. The protein concentration ranges between 15% and 20%.

The protein of the myofibrils is *inhomogeneous*. This lack of homogeneity causes the optical effect of cross-striation, that is, an alternating sequence of anisotropic (*A*) and isotropic (*I*) bands. The *A*-band of the myofibrils consists of *thick filaments* that are parallel and arranged in a hexagonal system. The *I*-bands consist of thin filaments, and are bisected by the *Z*-lines. The region between two *Z*-lines is called the *sarcomere* (Fig. 4-40).

The thick filaments are identical with the muscle protein myosin (about 38% of the total muscle protein), while the thin filaments contain mainly F-actin (14%) and tropomyosin. These proteins take part in the process of muscle contraction; therefore, they are called *contractile proteins*.

Six thin actin filaments are arranged around one thick filament. Both kinds of filaments are connected by *cross-bridges* (Figs. 4-41 and 4-43b).

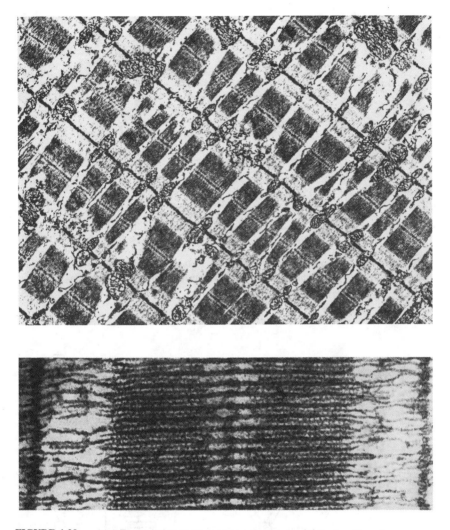

FIGURE 4-39.   (*top*) Transmission electron micrograph of a longitudinal section of rat *psoas* muscle at low magnification (10,000×). (*bottom*) High-resolution electron micrograph of a longitudinal section of glycerinated rabbit muscle (61,200×) (original provided by Dr. H. E. Huxley; with permission from Finean, 1967).

The Contractile Proteins

The myofibril of skeletal muscle (rabbit) consists of 21% actin, 54% myosin, 15% tropomyosin B, and 10% other proteins (α-actinin, β-actin, etc.).

After the water-soluble sarcoplasmic proteins are removed from the muscle tissue by extraction with a salt solution of low ionic strength ($\mu \sim 0.1$), the

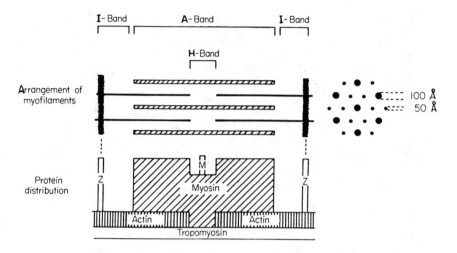

FIGURE 4-40.   Schematic representation of the arrangement of myofibrillar proteins within the *I*-, *A*-, and *H*-bands in muscle based on electron micrographs of muscle structure such as those illustrated in Figure 4-39.

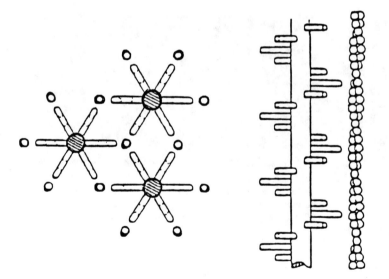

FIGURE 4-41.   Diagram to show the structure of the actin (thin) and myosin (thick) filaments. Note (*left*) the alignment of one actin filament, opposite each row of feet, in cross section; (*right*) the double-stranded beaded structure of actin, the pitch of the spiral being about 350°; and the six staggered rows of "feet" on the myosin filament.

contractile proteins of the myofibrils and the stroma proteins of the connective tissue remain. Extraction of this material by salt solutions of high ionic strength ($\mu \sim 0.6$) results in a strongly viscous solution that contains, besides myosin and actin, actomyosin (a complex of actin and myosin). Pure myosin can be obtained by stepwise lowering of the ionic strength.

Myosin

Myosin A is a thread-shaped molecule with a molecular weight (mol. wt.) of about 500,000. Myosin has enzymic properties: it breaks down adenosine triphosphate (ATP), the terminal phosphate group being split off, to yield inorganic phosphate and adenosine diphosphate (ADP), and releasing energy. This enzymic action of myosin is of great importance in the process of muscle contraction.

The adenosine triphosphatase (ATPase) activity of myosin is activated by calcium ions and inhibited by magnesium ions. Addition of actin changes the ATPase activity of myosin, which is very important to the muscle contractile system, because in the presence of actin, magnesium ions—at low ionic strength—have no inhibiting effect, but a strongly activating effect that is even greater than that of calcium ions. Here myosin ATPase is replaced by the actomyosin ATPase.

The myosin molecule is a composite system. It consists of at least two subunits that can be separated by tryptic digestion. According to their behavior in the ultracentrifuge, one unit, with a molecular weight of about 380,000, is called *heavy meromyosin* (H-meromyosin, HMM), and the other, with a molecular weight of about 120,000, is *light meromyosin* (L-meromyosin, LMM). Only the HMM shows ATPase activity, and only this part of the myosin molecule is responsible for the binding of actin. As electron micrographs show, myosin has a "head" formed by HMM and a "tail" formed by LMM (Fig. 4-42). The heads of the myosin molecules serve as cross-bridges connecting the thick and thin filaments in muscle (Fig. 4-43*b*).

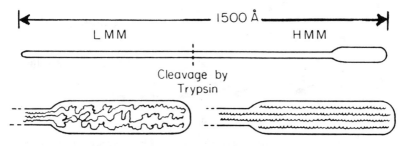

FIGURE 4-42.   Representation of the light and heavy meromyosin chains obtained after cleavage by trypsin.

Actin

Actin can be isolated from the material that is left after the partial extraction of myosin from muscle in the globular form (G-actin). The bead-shaped molecule of G-actin has a molecular weight of 50,000. To remain in the G-form some ATP must be present in the solution and bound to the actin. The G-actin polymerizes upon the addition of neutral salts to make chainlike molecules of fibrous actin (F-actin). This process is combined with simultaneous enzymic splitting of ATP into ADP and inorganic phosphate. In F-actin the globular particles are arranged side-by-side like the beads in a string of pearls (Fig. 4-43a and b).

Two of these strings are wound together to form a double helix. Although it has no definite length and therefore no definite molecular weight, it consists of several particles weighing together at least a million daltons.

Actomyosin

Mixing solutions of F-actin and myosin yields a solution of high viscosity that contains actomyosin. The addition of ATP, as well as inorganic polyphos-

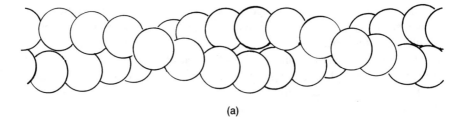

(a)

ACTIN FILAMENT

MYOSIN FILAMENT

CROSS-BRIDGES

(b)

FIGURE 4-43.   (a) Schematic representation of the double-helical (quaternary) structure of F-actin. (b) Model of the interaction between myosin and actin filaments in muscle through cross-bridge attachments.

phates, causes a drop in viscosity by dissociation of actomyosin into actin and myosin. When actomyosin is in the gel form, such as in myofibril, the addition of ATP under specified conditions causes the protein system to synerase ("superprecipitation"), or contract. In the latter case, pyrophosphate or tripolyphosphate are not able to simulate the action of ATP, which undergoes hydrolysis as syneresis, or contraction, occurs.

Actomyosin, synthesized by mixing pure myosin and actin (Fig. 4-43b), is not identical with actomyosin preparations isolated from muscle fibers (myosin B). Myosin B, or "natural actomyosin" consists of actin, myosin, tropomyosin, troponin, and α- and β-actinin.

Accessory Proteins

Tropomyosin and Troponin
Tropomyosin B represents a protein complex of tropomyosin and troponin. Tropomyosin shows no enzymic activity, nor does it combine with myosin. Tropomyosin probably is involved in the regulation of the interaction between myosin and actin (in the presence of ATP) in the process of muscle contraction (Fig. 4-44a and b).

Muscle Contraction and Rigor Mortis
The myofibrillar proteins actin, myosin, and tropomyosin, and troponin are of basic importance to the living organism because the contraction and relaxation of muscles are caused by their interactions. These proteins are also of particular interest for meat research, because of the drastic post mortem changes in meat. For example, the tenderness and ability to trap (or retain) water of meat are essentially due to changes in the actin–myosin system.

Contraction is induced by a release of very small amounts of calcium ions from the sarcoplasmic reticulum. These calcium ions activate myosin ATPase, causing a breakdown of ATP. As long as ATP is dephosphorylated, myosin and actin associate and form actomyosin. This transformation also happens after an animal is slaughtered. The stiffening of the muscle fiber post mortem, which occurs as the ATP level falls and as actin and myosin associate, is called *rigor mortis*.

Glycoproteins
Proteins containing covalently linked carbohydrate are called glycoproteins. These proteins are further classified as follows.

I. Proteoglycans
   A. Characterized by a very high carbohydrate content. Most mucopolysaccharides (MPS) in connective tissues are linked to protein

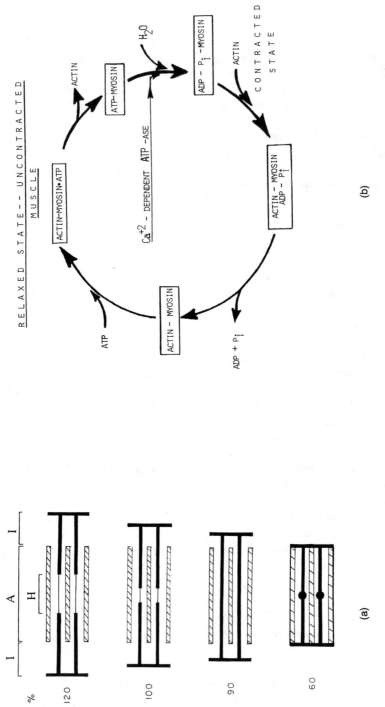

FIGURE 4-44. (*a*) Changes in the A-, H-, I-bands during muscle contraction. (*b*) Relationship between ATP and the cyclic association-dissociation of actin and myosin.

B. Characteristics of the carbohydrate component
   1. Carbohydrate units are large, containing more than 50 monosaccharide residues
   2. Most of the carbohydrate is present as a repeating disaccharide unit
   3. Hexosamine is always present, as is hexuronic acid (with one exception)

II. Glycoproteins
   A. The term *glycoprotein*, is generally used for a special protein, with relatively low carbohydrate content
   B. Characteristics of the carbohydrate component
      1. Carbohydrate units tend to be small, containing less than 25 monosaccharide residues
      2. The carbohydrate units have little or no repeating structure
      3. The types of monosaccharide present include fucose, mannose, galactose, glucose, *N*-acetylglucosamine, *N*-acetylgalactosamine, and sialic acid. Hexuronic acid is rarely, if ever, a component
   C. Structural glycoproteins (e.g., membrane glycoproteins, collagen)
   D. Mucus-type glycoproteins (e.g., mucins, blood group substances)
      1. Carbohydrate content is usually more than 50%
      2. The carbohydrate units are usually *O*-glycosidically linked to protein
      3. Of the monosaccharides just listed, the mannose content is usually low or nonexistent
      4. The protein component has a high content of threonine, serine, and proline
   E. Plasma-type glycoproteins (e.g., immunoglobulin, transferrin)
      1. Carbohydrate content is usually less than 50%
      2. The carbohydrate units are usually *N*-glycosidically linked to protein
      3. Of the monosaccharides just listed, the *N*-acetylgalactosamine content is usually low or nonexistent

Linkages between Carbohydrates and Proteins

Glycoproteins possess two types of linkages between protein and carbohydrate units, the *N*-glycosidic linkage and *O*-glycosidic linkage. Those glycoproteins containing *N*-glycosidic linkages are called *N*-glycosyl proteins, and those containing the *O*-glycosidic linkage are called *O*-glycosyl proteins. Most plasma-type glycoproteins belong to the *N*-glycosyl proteins, while proteoglycans, mucin-type glycoproteins, and collagen-type glycoproteins are *O*-glycosyl proteins.

1. In plasma-type glycoproteins, *N*-glycosidic linkages almost always occur between *N*-acetyl-*D*-glucosamine and asparagine. Furthermore, plasma-type glycoproteins invariably possess glycosamine residues in the form of di-*N*-

acetylchitobiose, a dimer of *N*-acetylglucosamine. These glycoproteins, however, lack *N*-acetyl-galactosamine.

These glycoproteins, however, lack *N*-acetyl-galactosamine.

2. In some other glycoproteins, *O*-glycosidic linkages occur between *N*-acetyl-*D*-galactosamine residue and the hydroxy group of serine (or threonine) residues.

3. Sometimes *O*-glycosidic linkages involve the hydroxyl group of hydroxyl-lysine residue of collagen.

Many glycoproteins contain *N*-acetylneuraminic acid.

The various protein structures illustrated in this chapter are stabilized by physical and chemical interactions that are illustrated in Figure 4-45*a* to *d*. A more detailed presentation of such interactions is given in Chapter 6 and in the second volume of this book.

NUCLEIC ACIDS AND THEIR COMPONENTS

Although nucleic acids are not food ingredients, they do control protein synthesis, which makes them important in food microbiology, genetic engineering, and biotechnology. They are polymers of four *nucleotides*, each of which consists of a base, a sugar ring, and a phosphate. The sugar component is either a *D*-ribose or a 2-deoxy-*D*-ribose (Fig. 4-46*a* and *b*, respectively) and determined if the nucleic acid is a ribonucleic acid (RNA) or a deoxyribonucleic acid (DNA).

The purine bases present in DNA are adenine and guanine, whereas the pyrimidine bases are thymine (5-methyl uracyl) and cytosine (Fig. 4-47*a*). Wheat germ DNA, for example, has a high proportion of *S*-methyl cytosine. In ribonucleic acids, on the other hand, the main pyrimidines are cytosine and uracil. The combination of a sugar component and a purine or pyrimidine nucleotide base is called a *nucleoside*. The structures of nucleotides, nucleosides, and polynucleotides are illustrated in Figure 4-47*a* to *d*, respectively. DNA structure was first derived from X-ray diffraction studies of fibers of a sodium salt of DNA (Watson and Crick, 1953a, b). Crick (1952, 1953), Watson (Watson and Crick, 1953a, b; Franklin and Gosling, 1953), and Wilkins et al. (1953) discovered the helical structure of DNA by comparing the *B*-DNA fiber diffraction pattern with the calculated diffraction pattern (Crick, 1953) of a helix (Fig. 4-48, page 164). Subsequently, a careful consideration of the H-bonding between purine–pyrimidine base pairs (Fig. 4-47*d*) and of the chemical nature of DNA components, in conjunction with detailed difffraction patterns of *A*-DNA crystals, resulted in a double-helical structure of DNA (Fig. 4-49, page 165), as proposed by Crick (Crick and Watson, 1956), Watson (Watson and Crick, 1953a, b), Franklin and Gosling (1953), Franklin (1956), and Wilkins (Wilkins, Stokes, and Wilson, 1953; Wilkins et al., 1953). Molecular models of the DNA double helix are presented in Figures 4-49 and 4-50*a* and *b* (page 166).

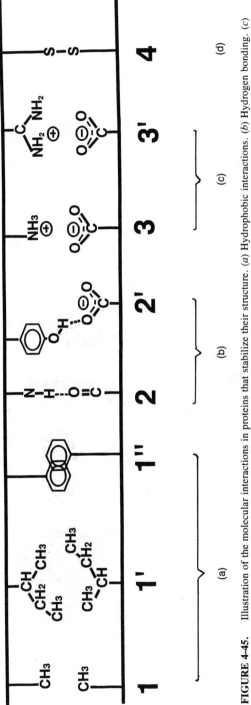

FIGURE 4-45. Illustration of the molecular interactions in proteins that stabilize their structure. (*a*) Hydrophobic interactions. (*b*) Hydrogen bonding. (*c*) Charge interactions. (*d*) Disulfide (*covalent*) bonding.

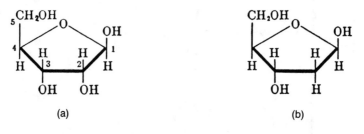

FIGURE 4-46.   Structural formula of (*a*) *D*-ribose, and (*b*) 2-deoxy-*D*-ribose.

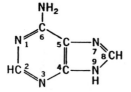

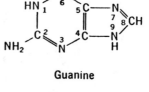

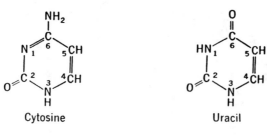

FIGURE 4-47*a*.   Structural formulas of some of the purines and pyrimidines found in nucleic acids.

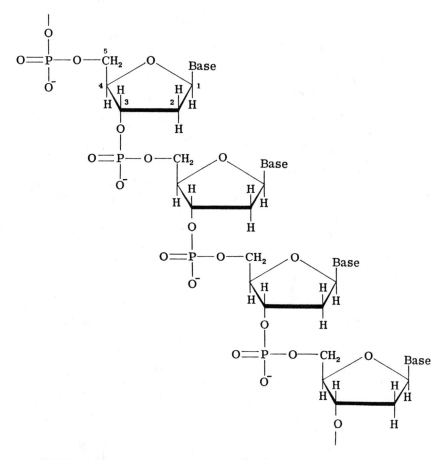

FIGURE 4-47b.   Structural formula of a segment of the deoxyribonucleic acid chain.

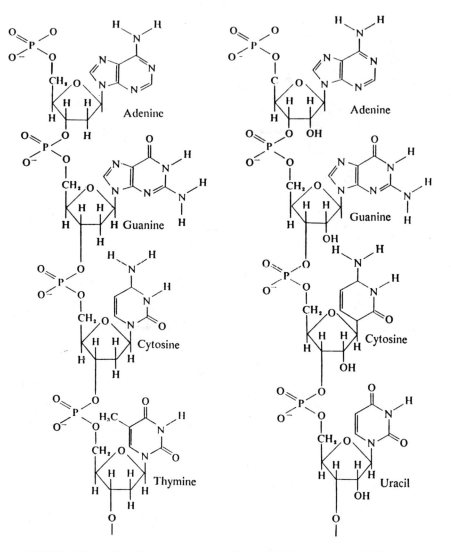

FIGURE 4-47c. Chemical structure of part of the (left) DNA chain, (right) RNA chain.

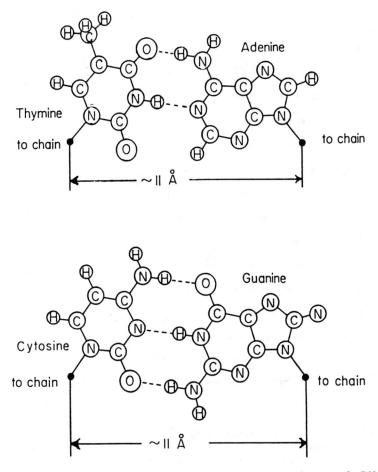

FIGURE 4-47d.   Detail of the specific pairing of bases in the proposed structure for DNA.

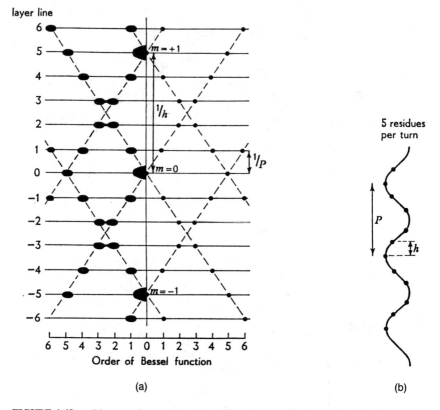

FIGURE 4-48.   Diagram of some of the Bessel functions contributing to the diffraction pattern (a) of a discontinuous helix (b) with five units per turn (shown at left; calculations by Crick, 1952, 1953) (from Wilson, 1966, with permission).

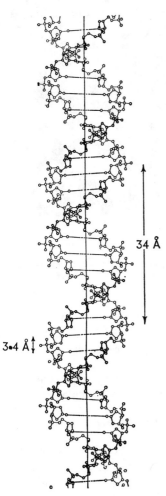

34 Å

3.4 Å

FIGURE 4-49.   The *B*-DNA structure. The two chains are antiparallel and are hydrogen bonded together through the bases (from Wilson, 1966, with permission).

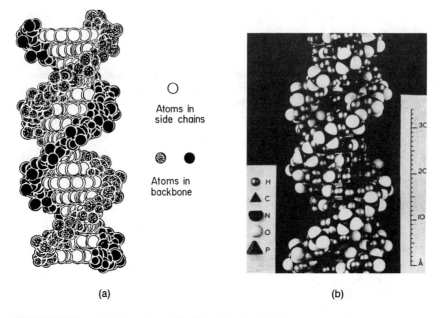

FIGURE 4-50.   (a) Schematic model of the DNA double-helix. (b) CPK molecular model of DNA (from Finean, 1967, with permission).

References

Baianu, I. C. 1981. Carbon-13 and proton NMR studies of wheat proteins in solution. *J. Sci. Food Agric.* **32**:309–313.

Baianu, I. C., L. F. Johnson, and D. K. Waddell. 1982. High-resolution, carbon-13 and nitrogen-15 nuclear magnetic resonance studies of wheat proteins at high magnetic fields: Spectral assignments, changes with concentration and heating treatments of *Flinor* gliadins in solution—Comparison with gluten spectra. *J. Sci. Food Agric.* **33**:373–383.

Conway, D. 1981. *Ionic Hydration in Chemistry and Biophysics.* New York: Elsevier.

Crick, F. H. C. 1952. Is α-keratin a coiled-coil? *Nature* **170**:882–883.

Crick, F. H. C. 1953. Fourier transform of a coiled-coil. *Acta Cryst.* **6**:685–687.

Crick, F. H. C., and J. D. Watson, 1956. Structure of small viruses. *Nature* **177**:473–476.

Finean, J. B. 1967. *Engström-Finean Biological Ultrastructure*, 2d ed. London and New York: Academic Press.

Franklin, R. E. 1956. Location of the ribonucleic acid in the tobacco mosaic virus. *Nature* **177**:928–930.

Franklin, R., and R. G. Gosling. 1953. Evidence for 2-chain helix in crystalline structure of sodium deoxyribonucleate. *Nature* **171**:156–157.

Lowrie, R. A., ed. 1970. *Proteins as Human Food*, Proceedings of the 16th Easter School in Agricultural Science, University of Nottingham, Nottingham, England. Westport, Conn.: AVI.

Prince, R. C., D. E. Ganson, G. S. Leight, and G. G. McDonald. 1982. The predominant form of fructose is a pyranose, not a furanose, ring. *Trans. Biochem. Soc.* **7**:239–241.

Schofield, J. D., and I. C. Baianu. 1982. Solid-state, cross-polarization magic-angle spinning carbon-13 nuclear magnetic resonance and biochemical characterization of wheat proteins. *Cereal Chem.* **59**(4):240–245.

Schulz, G. E., and R. H. Schirmer. 1978. *Principles of Protein Structure*. New York: Springer-Verlag.

Tokita, S. 1987. *Examples of Personal Computing*, Visual Quantum Chemistry Series (in Japanese). Tokyo: Kodan-sha Co.

Watson, J. D., and F. H. C. Crick. 1953a. Genetical implications of the structure of deoxyribonucleic acid. *Nature* **171**:964–967.

Watson, J. D., and F. H. C. Crick. 1953b. Molecular structure of nucleic acids: A structure for deoxypentose nucleic acids. *Nature* **171**:737–738.

Whitney, R. McL. 1977. Chemistry of colloidal substances: General principles. In *Food Colloids*, H. D. Graham, ed., 14–21. Westport, Conn.: AVI.

Wilkins, M. H. F., A. R. Stokes, and H. R. Wilson. 1953. Molecular structure of nucleic acids: Molecular structure of deoxypentose nucleic acids. *Nature* **171**:738–739.

Wilkins, M. H. F., W. E. Seeds, A. R. Stokes, and H. R. Wilson. 1953. Helical structure of crystalline deoxypentose nucleic acid. *Nature* **172**:759–762.

Wilson, H. R. 1966. *Diffraction of X-rays by Proteins, Nucleic Acids and Viruses*. London: Edward Arnold.

Zubay, G. L. 1983. *Biochemistry*. Reading, Mass.: Addison-Wesley.

5

Techniques

An increasing number of physical and chemical techniques are being applied to study the complex properties of biopolymers, foods, and their components. Traditionally, the emphasis has been placed on *rheological* measurements in relation to the "functionality" of foods. Such techniques are expanded upon in Chapter 8, in the context of testing cheese; other similar applications of rheological techniques are further discussed in Chapter 9 in the context of food extrusion.

The techniques presented briefly in this chapter are from three major groups that are well established.

1. Structural techniques: microscopy, diffraction, and scattering
2. Spectroscopic techniques: infrared absorption, Raman scattering, photoacoustic spectroscopy (PAS), nuclear magnetic resonance (NMR), and electron spin resonance (ESR)
3. Thermoanalytical techniques: differential scanning thermal calorimetry (DSC), differential thermal analysis (DTA), and thermomechanical analysis (TMA)

The first group is concerned mostly with scattering or diffraction processes involving light, X-rays, electrons, or neutrons; the second group is concerned with the observation of absorption of electromagnetic radiation, and the third group deals with heat absorption/emission processes. At least one technique from each group is covered here, while other related techniques are looked at in greater detail, with food applications, in Volume II of this book.

STRUCTURAL TECHNIQUES

X-ray Diffraction

The *structure* of ordered solids or crystals is routinely determined by X-ray diffraction (XD). As the name indicates, one employs a focused X-ray beam to obtain an X-ray diffraction pattern of the crystal or crystalline powder under investigation. The reason for routinely using X-rays rather than neutrons is the relative ease and low cost of generating X-rays by bombarding a metal strip or rod with accelerated electrons. A narrow range of X-ray radiation wavelengths is selected by a crystal *monochromator*, so that the experimenter can readily calculate interatomic distances in the material under study from the known wavelength of the X-rays and the geometric characteristics of the diffraction camera. A typical setup for fiber X-ray diffraction studies is shown in Figure 5-1a to c. Such an experimental arrangement is suitable for studying collagen, any other ordered protein fibers, oriented DNA (or RNA) fibers, viruses, biological membranes, and lipid or polysaccharide films.

The principle of the X-ray diffraction technique can be briefly summarized as follows. The incident X-ray beam is scattered, or diffracted, by the regular crystal lattice (Fig. 5-1d) following relatively simple diffraction laws. The scattered waves from neighboring atoms interfere with each other, resulting in an interference or diffraction pattern of alternating high and low intensity. A photographic plate, or a sensitive detector, is exposed to the diffracted waves and records their diffraction pattern in a permanent form (for example, on a film or in a photograph). Such a diffraction pattern consists of a large number of diffraction spots aligned along the directions for which constructive reinforcement of the scattered X-rays occurred. *Bragg's law*, which relates the diffraction pattern to the crystal structure, is often stated simply as follows: The scattered X-ray beam behaves as if it was a reflection of the incident X-ray beam into the "mirror," formed by the lattice planes of the crystal (Fig. 5-2a and b). The corresponding *Bragg equation* gives the positions of the Bragg reflections stated in terms of the scattering (or diffraction) Bragg angle θ in Figure 5-2a as a function of the spacing between the lattice planes d

$$n\lambda = 2d \cdot \sin \theta \qquad (5\text{-}1)$$

where n is an integer, the *order* of the diffracted beam, and λ is the wavelength of the incident X-rays.

A well-defined X-ray diffraction pattern is obtained only if the regions of crystalline structure have a certain minimum size (approximately 100 to 200 Å), so that the X-ray interference of the scattered waves is repeated many times. Furthermore, the "reflections" in the diffraction pattern correspond to

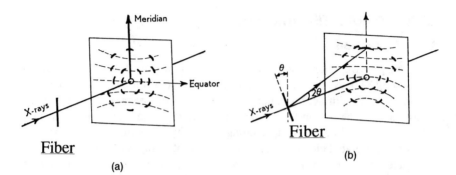

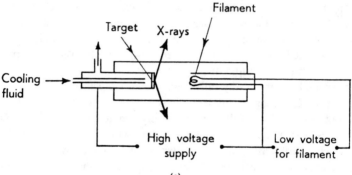

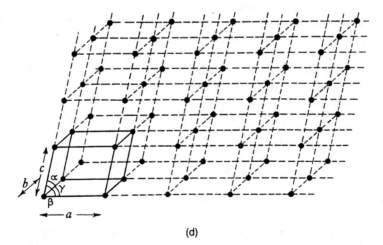

FIGURE 5-1.   (*a*) Fiber diffraction: the fiber is perpendicular to the X-ray beam. (*b*) The fiber is tilted at the appropriate angle to record a meridional reflection. (*c*) X-ray generator operating with a sealed tube (modified from Wilson, 1966). (*d*) A diagram representing a crystal lattice. One *unit cell* is outlined with a continuous line.

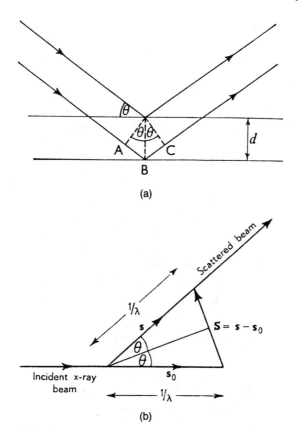

(a)

(b)

FIGURE 5-2.   (a) Illustration of *Bragg's law* of X-ray diffraction ($n\lambda = 2d \sin \theta$). (b) Vector representation of the scattered beam. In this figure, d is the distance between the lattice plane, and θ is the Bragg angle made by the incident beam with the normal to the lattice planes.

maxima in the resulting scattered wave that have time differences with respect to each other, called *phases*. In order to calculate the three-dimensional structure of the crystal from its X-ray diffraction pattern we need both the intensities of the scattered beam and phases. Unfortunately, the phase information is lost during the detection of the scattered X-rays, and so they have to be found either from model-building considerations or by other sophisticated methods. In special cases, the phases can be determined by placing heavy metal atom "phase markers" in the crystal, without major alterations to the crystal structure. One can find the phases of the X-ray reflections from the difference in the X-ray diffraction patterns, with or without the heavy metal marker, and then use it to calculate the complete three-dimensional crystal structure. A simple example of an X-ray diffraction pattern is given in Figure 5-3a for the simple cubic lattice

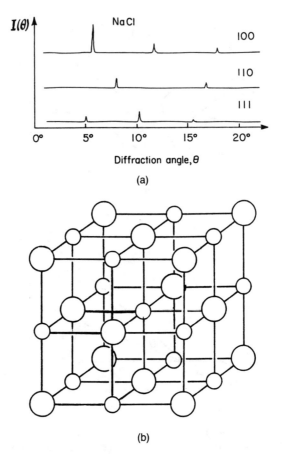

FIGURE 5-3.   (a) Diffraction pattern for NaCl, for the [100], [110], and [111] directions. (b) Schematic representation of the cubic lattice of NaCl (both (a) and (b) from Brey, 1972).

(Fig. 5-3b) of the common salt NaCl. A crystalline powder can also be studied by X-ray diffraction to obtain some of the characteristics of the crystal lattice. Because the crystal lattice structure is periodic (repeats itself), it can be mathematically expressed as a sum of sine or cosine terms, called a *Fourier series*. The electron density distribution, ρ, varies periodically throughout the crystal lattice, and the same pattern is repeated for each *unit cell* of the lattice. Along a specified direction, the electron-density distribution in the crystal lattice can therefore be expressed as the Fourier series:

$$\rho(x) = \sum_{h=-\infty}^{h=+\infty} C_h \cdot \cos 2\pi \left(\frac{hx}{a} + \phi_h \right) \tag{5-2}$$

where h is the order in the series, C_h are coefficients that need to be determined, x is the lattice point at which $\rho(x)$ is to be calculated, a is the repeat distance in the lattice along one of the three axes of the unit cell, and ϕ_h is the phase angle, fixed by the value of this function at the chosen origin, $x = 0$.

An artificial, *one-dimensional* lattice with atoms distributed at regular intervals along a single line would give a one-dimensional X-ray diffraction pattern corresponding to the electron density in equation (5-2). The intensity of an X-ray reflection or diffraction spot is then found to be proportional to the square of the coefficients C_h in equation (5-2). The latter are therefore called *structure factors* and are denoted by F_h. The structure factor sums up the scattering effects of all the atoms in the unit cell, and could be calculated for a given structure by superimposing all the scattered beams from the various atoms with suitable phase differences. For the single one-dimensional case considered here

$$F_h = \sum_i F_i \cdot \cos 2\pi \left[\frac{hx_i}{a} + \phi_i(h) \right] \qquad (5\text{-}3)$$

where i labels the particular coordinate x_i for which a term F_h is being evaluated, F_i is the scattering power of the atom i, and a is the repeat distance in the one-dimensional lattice. The phase angle ϕ_i is determined by the distance of the atom i from the point in the unit cell that was chosen as the origin ($x = 0$). An example of such a one-dimensional X-ray diffraction pattern from oriented erythrocyte membranes is presented in Figure 5-4, and is further analyzed in Chapter 17, Volume II of this book.

For a periodic three-dimensional lattice the structure repeats in three directions, **a**, **b**, and **c**, with repeat distances (unit cell dimensions), a, b, and c. In the latter case, the structure factors have a form similar to the simpler one-dimensional lattice previously considered (eq. (5-3)), except that three sets of indices, h, k, and l, are needed to specify the lattice planes. Thus, the structure factors of a crystal lattice are

$$F(h, k, l) = \sum_i F_i \cdot \cos 2\pi \left[\frac{hx_i}{a} + \frac{ky_i}{b} + \frac{lz_i}{c} + \phi(h, k, l) \right] \qquad (5\text{-}4)$$

The electron-density distribution in the crystal lattice can then be expressed as

$$\rho(x, y, z) = \frac{l}{V_o} \sum_h \sum_k \sum_l F(h, k, l) \cdot \cos 2\pi \left[\frac{hx}{a} + \frac{ky}{b} + \frac{lz}{c} + \phi(h, k, l) \right]$$

$$(5\text{-}5)$$

FIGURE 5-4. One-dimensional, low-angle X-ray diffraction pattern for a partially disordered lattice of a system of parallel erythrocyte membranes close packed into a film.

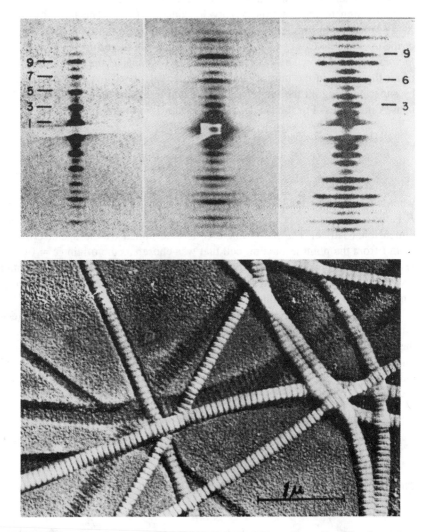

FIGURE 5-5. *Top:* Low-angle X-ray diffraction pattern from collagen fibers showing more than nine Bragg orders (n) of diffraction. *Bottom:* Electron micrograph of shadowed collagen fibrils whose X-ray diffraction pattern is shown above. (Both illustrations from Finean 1967, with permission.)

where V_o is the volume of the unit cell. This equation therefore allows the three-dimensional structure of any crystal to be calculated, if the phase angles ϕ are known, by determining the amplitude of the scattered X-ray beam from the X-ray diffraction pattern of the crystal.

Oriented fibers, for example, give characteristic diffraction patterns that are readily recognized. This pattern is illustrated in Figure 5-5a and b for fibrous protein, and in Figure 5-6a, b, and c for a DNA crystal consisting of regularly packed double helices. Various polysaccharides can be oriented in films or fi-

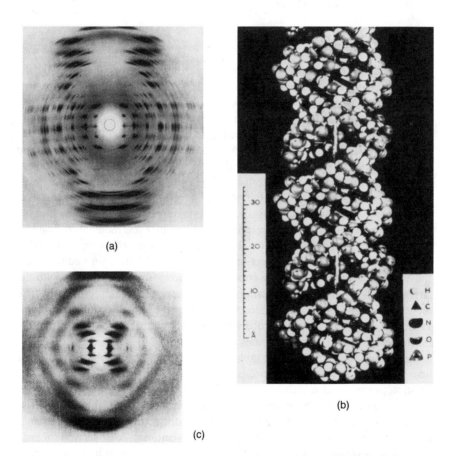

(a)

(b)

(c)

FIGURE 5-6.   (a) X-ray diffraction pattern of Na DNA in the A conformation at 75% relative humidity (original photograph courtesy of Dr. M. H. F. Wilkins; Wilkins et al., 1953; Watson and Crick, 1953b; Franklin and Gosling, 1953; Franklin, 1956). (b) CPK molecular model of A-DNA structure. (c) X-ray diffraction photograph of Na DNA in the B conformation at 92% relative humidity (original photograph courtesy of Dr. M. H. F. Wilkins; Wilkins, Stokes, and Wilson, 1953; Wilson, 1966).

bers, and their structure could be determined by this technique. Depending on the source and their degree of hydration and crystallinity, starch granules give several types of X-ray diffraction patterns labeled *A, B, C, V* (Zobel, 1988). Such structural information is valuable to the food industry, both for quality control and food product development. In these more complex cases, however, *complete* structural information may be unavailable from X-ray diffraction alone. In that case, spectroscopic techniques, such as solid-state NMR, may provide valuable, complementary information (further details of such techniques will be given in Vol. II of this book).

Neutron Diffraction

Whereas X-ray diffraction is concerned with the scattering of X-rays by the outer molecular or atomic (electron) orbitals in a crystal lattice, neutron diffraction (ND) is caused by the interactions of an incident neutron beam with the nuclei in a scattering target. The neutron scattering amplitude is strongly dependent on the nature of the scattering nuclei. Neutron diffractometers are quite large instruments compared with routine X-ray diffractometers (excluding those X-ray machines exploiting a synchrotron source of X-rays). Neutrons generated in a nuclear reactor are collimated to an incident beam that is scattered by a relatively large crystal (of the order of several millimeters), and the scattered neutron beam is detected by a bank of counters covering the angular range from about 0.1° to 120°, corresponding to a range of momentum transfer $(hk/2\pi)$ from 0 to above 11 Å^{-1} for thermal neutrons of about 1-Å wavelength. Scattering intensity corrections are far more elaborate for neutrons than for X-rays, because of the more complex interactions of the neutrons with the target, involving multiple scattering, inelastic scattering, and absorption. The reduced scattered intensity is scaled by means of the vanadium standard, since vanadium gives rise to only incoherent neutron scattering. Neutron diffraction is often employed to refine structures already obtained by X-ray diffraction techniques. However, a much wider range of experiments is possible with neutrons than with X-rays; among these are isotope labeling experiments, magnetic orientation processes in metals, and dynamic processes in solids and liquids. Because hydrogen atoms have very low X-ray scattering power, they are "invisible" in the structures reconstructed by X-ray diffraction. On the other hand, hydrogen atoms have an appreciable incoherent scattering cross section for neutrons, and their positions can be very accurately determined in large crystals by neutron diffraction. The large mass of the neutrons gives a high probability for their interaction with hydrogen atoms.

An example of the accuracy obtainable in crystal structure determination by neutron diffraction is presented in Figure 5-7, where the bond distances in the sucrose molecule were determined with an accuracy of ± 0.002 Å (or better)

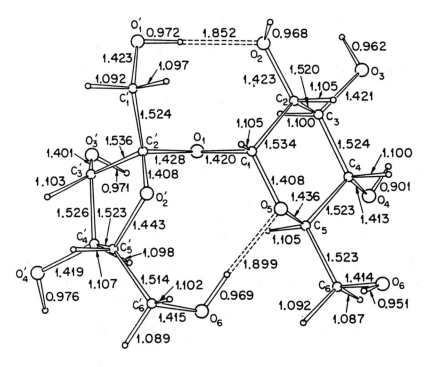

FIGURE 5-7.   The sucrose molecule, showing highly precise interatomic distances, as determined by neutron diffraction. The small circles are hydrogen atoms, and the dashed lines indicate hydrogen bonds from the hydroxylic hydrogens to neighboring oxygens (courtesy of Dr. G. M. Brown and Dr. H. A. Levy, Oak Ridge National Laboratory).

and the bond distances for hydrogens were accurate to ± 0.005 Å or better (Brey, 1972). Neutron diffraction has also provided the most accurate representation of *local* structure in liquid water (D_2O) (Soper and Phillips, 1986), although not all the details of the analysis are universally accepted (Baianu et al., 1992). In difficult systems, such as liquid water, even a combination of X-ray, neutron, and electron diffraction data is insufficient to obtain all the structural details. In this case, spectroscopic techniques such as NMR and Raman scattering provide valuable dynamic and structural information, complementary to that obtained by scattering techniques (see also the section on liquid water structure in Chap. 6).

The calculation of the distribution of nuclei in a crystal lattice and of the coherent elastic scattering of neutrons by a crystal proceeds in a manner very similar to that employed by X-ray diffraction. However, the neutron scattering intensity corrections are much more difficult than in the X-ray diffraction studies, and the inelastic scattering correction is a major source of uncertainty for the more complex liquids (for example, liquid water (H_2O)). Studies of liquid water by neutron diffraction are mostly carried out on heavy water (D_2O), because the deuteron scattering of neutrons is *coherent*, unlike that of hydrogens in light water (H_2O). Furthermore, the scattering theory for liquid water involves elaborate derivations that are beyond the scope of this introductory text (for additional details, see the recent review by Baianu et al., 1992).

Microscopy

One of the major reasons for the wide popularity of microscopy in biology and material sciences is that the diffracted light, or diffracted electron beam, is converted directly *by physical means* into a reconstructed picture or magnified image of the object structure, often as a projection rather than a three-dimensional picture. The phase angles are not lost in the detection process (as they are in the use of the scattering/diffraction techniques described in the previous section); instead, the reconstruction of the magnified image of the object structure is obtained by a physical device such as a projector lens. For routine microscopic observations one does not therefore require either a detailed understanding of the theory of image reconstruction or the use of elaborate calculations with a fast digital computer. Despite its apparent simplicity, microscopy involves *Fourier transformation* (FT), as much as X-ray, or electron, diffraction analysis. However, the FT is implemented by a physical device, such as a lens, rather than being carried out digitally. There is, however, no such lens available for X-rays. In the case of electron microscopy the lenses are either magnetic, electromagnetic, or electrostatic devices. A schematic diagram of a transmission electron microscope (TEM) is shown in Figure 5-8a. Newer techniques, such as scanning electron microscopy (SEM or STEM), as well as a series of other recently developed EM techniques, involve more complex processes (Fig. 5-8b) than those used in transmission electron microscopy, and rather sophisticated principles of operation. Additional, very detailed structural information is currently obtainable through the latter techniques. One of the major limitations of most EM techniques is the way hydrated samples or liquids need to be handled. Since the sample has to be placed in a high-vacuum chamber for examination, the sample is very dry, frozen, or coated. Another disadvantage of TEM is the necessity of staining and fixation for biological/food samples because of the poor contrast at the low beam intensities that are employed to avoid damage to the specimen. Such treatments alter the sample structure and can cause artifacts. In favorable cases, high-quality images can be obtained without staining or shadowing (Fig. 5-9), or through digital FT of an electron diffraction image.

An illustration of the TEM technique is given in Figure 5-10, which shows the arrangement of collagen fibrils in the stroma of chick cornea. In areas of

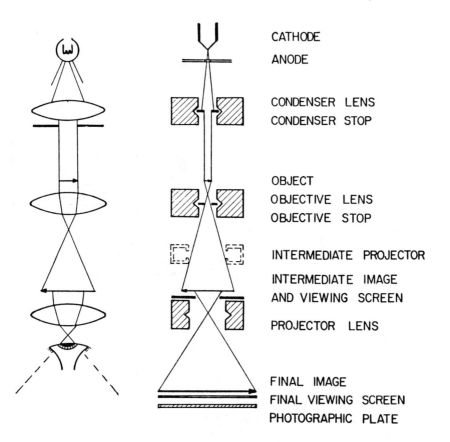

CATHODE

ANODE

CONDENSER LENS
CONDENSER STOP

OBJECT
OBJECTIVE LENS
OBJECTIVE STOP

INTERMEDIATE PROJECTOR

INTERMEDIATE IMAGE
AND VIEWING SCREEN

PROJECTOR LENS

FINAL IMAGE
FINAL VIEWING SCREEN
PHOTOGRAPHIC PLATE

FIGURE 5-8a.   Diagram illustrating the layout of an electron microscope in comparison with the light microscope.

food science, such as *food microbiology*, microscopic techniques are indispensable. There are also numerous applications of EM techniques in quality control and product development in the food industry. Often such applications require a certain degree of familiarity with the specific characteristics and problems associated with the handling of foods. The structural information readily obtained by EM techniques (Figs. 5-11 to 5-14, on pages 181–184) for food samples, or materials, often cannot be gotten by X-ray diffraction or other techniques, primarily because of the complexity of the food systems. X-ray diffraction techniques are often looked upon as a refinement, or elaboration, on the structure obtained by an EM technique. The combination of EM, or SEM, with X-ray fluorescence provides a powerful method for chemical analysis of biological and food samples. A food-specific application is the sensitive detection

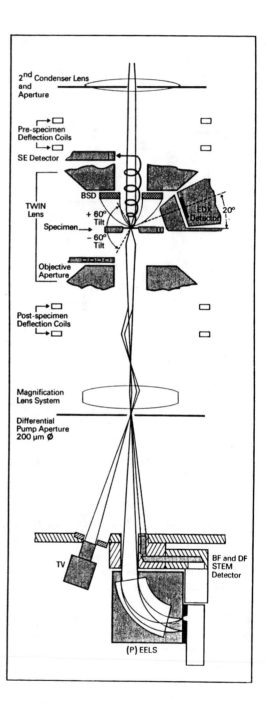

FIGURE 5-8*b*.   Detector configuration for the STEM showing secondary-electron, backscattered-electron, and energy-dispersive X-ray detectors in the specimen environment and STEM bright-field/dark-field detectors (near axis), EELS (on-axis), and CCD TV (off-axis) mounted simultaneously below the projection chamber. For the last three cases, the microscope automatically deflects the beam to the detector required when the fluorescent screen is lifted.

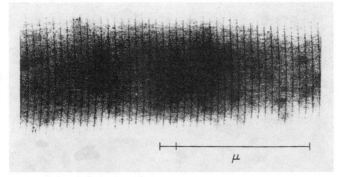

FIGURE 5-9.   Electron micrograph of an unstained and unshadowed preparation of L-meromyosin (courtesy of Dr. Philpott and Dr. Szent-Görgyi; with permission from Finean, 1967).

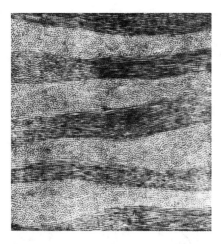

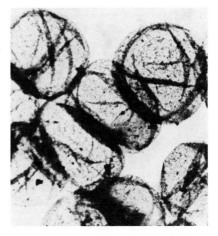

FIGURE 5-10.   Electron micrograph showing arrangement of collagen fibrils in the stroma of chick cornea (illustration provided by Dr. Marie A. Jakus; with permission from Finean, 1967).

FIGURE 5-11.   Electron micrograph of negatively stained human erythrocyte ghosts under isotonic conditions (Baianu, 1974).

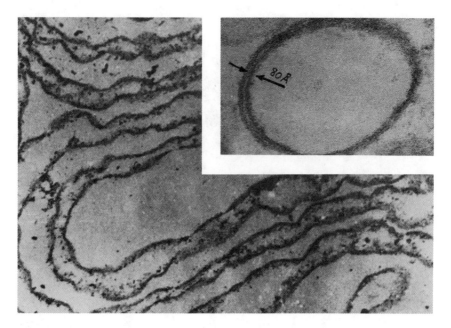

FIGURE 5-12. Transmission electron micrograph of thin sections of OsO_4 stained and fixed human erythrocyte membranes. The inset shows ~ 80-Å-thick membranes in the dehydrated state, after glutaraldehyde fixation and OsO_4 staining (Baianu, 1974).

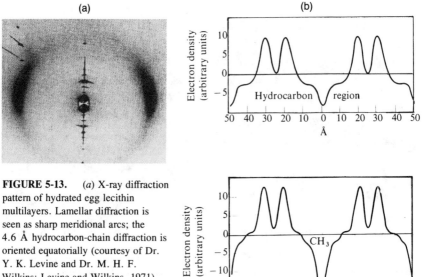

(a)

(b)

FIGURE 5-13. (a) X-ray diffraction pattern of hydrated egg lecithin multilayers. Lamellar diffraction is seen as sharp meridional arcs; the 4.6 Å hydrocarbon-chain diffraction is oriented equatorially (courtesy of Dr. Y. K. Levine and Dr. M. H. F. Wilkins; Levine and Wilkins, 1971). (b) Fourier synthesis across egg lecithin bilayers at 57% relative humidity (after Levine and Wilkins, 1971).

182

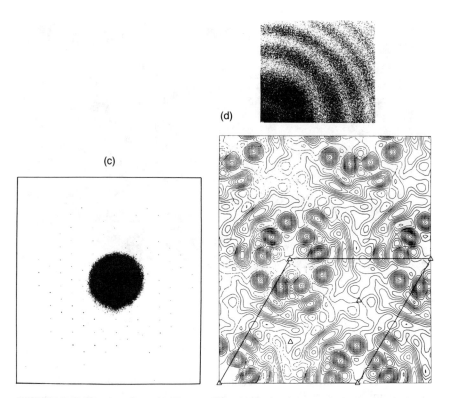

(d)

(c)

FIGURE 5-13 (Continued).   (c) Electron diffraction in the electron microscope for purple membranes. (d) Structure of purple membranes derived from electron diffraction in the electron microscope. (Inset shows higher electron diffraction orders.) Electron density map at 7 Å resolution, calculated by a Fourier synthesis using the measured intensities and calculated phases. The unit cell dimensions are 62×62 Å. (Both (c) and (d) courtesy of Dr. P. N. T. Unwin and Dr. R. Henderson; with permission from Cantor and Shimmel, 1980.)

and determination of the distribution of undesirable, toxic (or nontoxic), metallic impurities in some food products.

A simple yet detailed description of a wide range of the biological applications of EM techniques (as well as some X-ray diffraction applications) is available in textbook form with numerous high-quality illustrations (Finean, 1967).

SPECTROSCOPIC TECHNIQUES

Spectroscopy can be defined, in general, as the study of the interactions of electromagnetic radiation with atoms and molecules. Spectral analyses can be employed to obtain information about the chemical composition of samples,

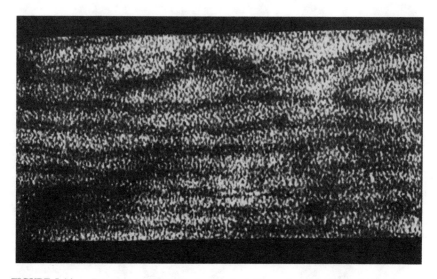

FIGURE 5-14.   Transmission electron micrograph of thin-sectioned fixed and OsO_4 stained samples of oriented erythrocyte membranes.

molecular structure, and dynamics. In addition, the kinetics of various processes can be determined by special adaptations of spectroscopic techniques. Because of their relative ease of use, spectroscopic determinations are very popular. Chemical composition analysis and kinetic measurements are beyond the scope of this chapter; the former are routinely handled by analytical laboratories. The foundation for spectroscopic measurements is *quantum mechanics* (which was briefly presented in Chap. 3). The electromagnetic (e.m.) spectrum covers a very wide range of frequencies, from radio waves ($\sim 3 \times 10^5$ Hz) to gamma rays (3×10^{21} Hz). Table 5-1 lists the type of electromagnetic radiation commonly encountered, their corresponding frequencies, wavelengths (λ), and wave number ($1/\lambda$). Most spectroscopic methods are concerned with the absorption of radiation as a function of frequency (or wavelength) in one of the ranges specified in Table 5-1. A *spectrophotometer* is an instrument that measures the absorption of photons in a sample; spectrophotometers are specifically built to measure separately the ultraviolet/visible and infrared ranges of photon absorption. A schematic diagram of a spectrophotometer is shown in Figure 5-15, where only the main components are indicated. Not shown in Figure 5-15 is a microcomputer or plotter/printer that receives and records the output from the detector for permanent storage and subsequent data analysis. The emission of photons by an illuminated sample is of great interest to researchers, especially in biochemical areas, because of the very high sensitivity of such fluorescence techniques.

TABLE 5-1   The Electromagnetic Spectrum

Type of Radiation	Wavelength			Frequency (cycles/sec or Hz)	Wave Number (cm^{-1})
	cm	microns	Å		
Radio	10^5			3×10^5	10^{-5}
Microwaves	10^1			3×10^9	10^{-1}
Far infrared	10^{-1}	1,000		3×10^{11}	10^1
Near infrared	2×10^{-3}	20	200,000	1.5×10^{13}	5×10^2
Visible	7.5×10^{-5}	0.75	7,500	4×10^{14}	1.3×10^4
Near ultraviolet	4×10^{-5}	0.4	4,000	7.5×10^{14}	2.5×10^4
Vacuum ultraviolet	2×10^{-5}	0.2	2,000	1.5×10^{15}	5×10^4
X-rays	2×10^{-7}		20	1.5×10^{17}	5×10^6
Gamma rays	10^{-9}		0.1	3×10^{19}	10^9
	10^{-11}		0.001	3×10^{21}	10^{11}

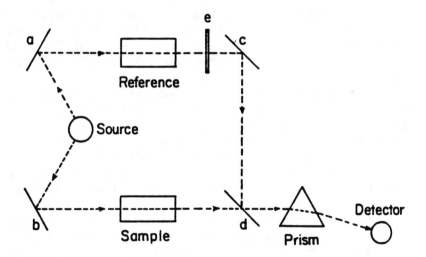

FIGURE 5-15.   Schematic diagram of a spectrophotometer. Radiation is reflected from the source through the sample and through a reference by mirrors. The reference beam is attenuated by e, which is a wedge that can be moved into and out of the light path, until it is equal in intensity to the sample beam. The two beams are alternately sent to the detector by d, which may be a rotating semicircular mirror. A grating, instead of a prism, is frequently used as a monochromator.

The resonant absorption of radio waves and/or microwaves by a sample situated in a magnetic field is known as *magnetic resonance* (MR). The resonant absorption of microwaves by the *unpaired* electron spin (such as those of radical species, or paramagnetic materials) in a magnetic field has become known as *electron spin resonance* (ESR) or *electron paramagnetic resonance* (EPR). The resonant absorption of radiowaves by the nuclear spins in a liquid, solid, or gaseous sample placed in a magnetic field is known as *nuclear magnetic resonance* (NMR). Within the NMR field, four areas of specialization have emerged: high-resolution/two-dimensional NMR (HR-NMR), nuclear magnetic relaxation (or nuclear spin relaxation), solid-state NMR, and NMR imaging (or magnetic resonance imaging (MRI)/zeugmatography). Of these techniques, high-resolution NMR is the most widely used (especially in chemistry/biochemistry laboratories), while MRI is being increasingly used in the medical field (so that the letter "M" in the technique name might soon stand for "medical"). The HR-NMR technique is so popular *not* because of its high sensitivity (which could not be claimed) but because of its very high resolution of individual nuclei in sample solutions, as its name adequately indicates. Two-dimensional NMR adds even more resolution and sophistication to an already powerful and convenient technique.

The instruments that do not measure light absorption are called *spectrometers* to distinguish them from spectrophotometers (which measure only photon absorption). Thus, there are ESR, NMR, mass, and Mössbauer (gamma) spectrometers, all of which have been specially designed to measure the corresponding e.m. absorption processes. (Further instrumentation details are presented in the second volume of this book, in Chap. 15).

Two of the spectroscopic techniques discussed previously, NMR and ESR, are presented in further detail in the next section (and in Vol. II). From an instrumental standpoint one can distinguish two groups of spectrometers and spectrophotometers: one that employs continuous wave (c.w.) excitation and detection, and one that employs short pulses of radiation for excitation, as well as Fourier transformation, after the signal detection. The second group requires the use of an *on-line computer* or array processor to carry out the FT, and is generally more powerful and sophisticated than the first group. Pulsed, FT-NMR, and FT-IR spectrometers are in the second group. Pulsed FT-EPR spectrometers are not as widespread as pulsed, FT-NMR, or FT-IR spectrometers; most EPR spectrometers are currently operated in the c.w. mode, whereas there are only a few c.w. NMR spectrometers still in use.

Nuclear Magnetic Resonance Spectroscopy and Relaxation

Nuclear magnetic resonance (NMR) is a branch of radio-frequency (r.f.) absorption spectroscopy that is specifically concerned with the *resonant* absorp-

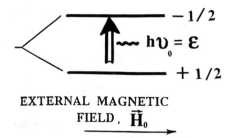

EXTERNAL MAGNETIC
FIELD, $\vec{H}_0$

FIGURE 5-16.   The resonance condition in NMR, $h\nu_0 = g\beta_N I \cdot H_0$, where $g\beta_N \cdot I = \gamma_N$ is the magnetogyric ratio of the nucleus.

tion of radio waves by the nuclei of a sample placed in an intense magnetic field. The radio-frequency absorption occurs as a result of transitions between the nuclear spin energy levels, in the presence of a very homogeneous, strong, and static magnetic field. The stronger the magnetic field, $\vec{H}_0$, the higher the frequency required for resonance, and the more intense the absorption, as indicated in Figure 5-16, and Table 5-2.

The transition induced between the nuclear spin energy levels (Fig. 5-16) by the resonant r.f. wave or pulse is recorded as a sharp peak for a liquid; for the pair of nuclear spin energy levels shown in Figure 5-16, a single peak is recorded whose lineshape is Lorentzian in a simple liquid or solution. The linewidth of the NMR absorption peak at half-height is determined by the lifetime of the excited nuclear spin state in the presence of only negligible magnetic field inhomogeneities. More precisely expressed, after the occurrence of the NMR absorption, the system of nuclear spins relaxes through interactions between the nuclear spins (spin–spin, or T_2-relaxation process). The NMR signal is recorded with a coil whose axis, x, is perpendicular to the direction, z, of the

TABLE 5-2   **NMR Characteristics of Several Isotopes Encountered in Biological Systems and Foods**

Isotope	NMR Frequency (MHz) at 14.092 kG	NMR Frequency (MHz) at 23.487 kG	Natural Abundance (%)	Relative Sensitivity for Equal Number of Nuclei	Nuclear Spin
^1H	60.00	100.00	99.985	1.00	1/2
^2H	9.210	15.351	0.015	9.65×10^{-3}	1
^{13}C	15.087	25.144	1.108	1.59×10^{-2}	1/2
^{14}N	4.334	7.2238	99.63	1.01×10^{-3}	1
^{17}O	8.134	13.56	0.037	2.91×10^{-2}	5/2
^{19}F	56.446	94.077	100.0	0.833	1/2
^{31}P	24.288	40.481	100.0	0.066	1/2

static magnetic field, $\vec{H}_0$, in Figure 5-16; therefore, the loss of phase coherence of the nuclear spins in the xy-plane, which is normal to $\vec{H}_0$, occurs as a result of transverse (T_2) nuclear spin relaxation processes (Fig. 5-17a) that involve interactions between nuclear spins. The corresponding free induction decay (FID) signal is shown in Figure 5-17b. On the other hand, the nuclear spin magnetization has a component M_z along the magnetic field direction (z), which is not directly observed by the detector coil whose axis is in the xy-plane, along the x-direction. Immediately after the nuclear spin excitation, the M_z component points against the static magnetic field $\vec{H}_0$ and later relaxes (or comes back) with a characteristic time constant, T_1, toward the magnetic field direction. This

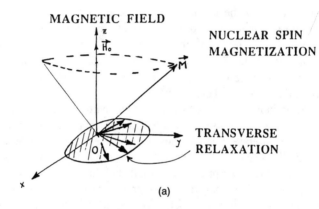

(a)

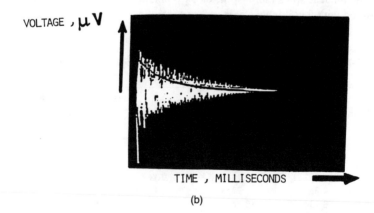

(b)

FIGURE 5-17.   (a) Precession of nuclear spins and evolution of the transverse nuclear spin magnetization component in a T_2, NMR measurement. (b) The free induction decay NMR signal (FID) corresponding to the decay of the *transverse* magnetization components in part (a).

relaxation process, which occurs along the z-axis, is called the *longitudinal*, or *spin-lattice* (T_1) relaxation. The latter name is often used because the T_1-relaxation process involves interactions of the nuclear spins with the surrounding electrons (the lattice). In a crystalline solid, the nuclear spins interact with the electrons in the surrounding crystal lattice, causing the relaxation of the M_z magnetization component toward the magnetic field direction. The use of the term "spin-lattice" relaxation is, however, not restricted to T_1-relaxation in crystalline solids, but is employed in general for any system, either solid, liquid, or gas.

Mechanisms of Relaxation in a Liquid

For a spin-$1/2$ (such as ^1H) the major contributions to relaxation are made by spin–spin coupling, magnetic dipolar interactions, and chemical exchange (of protons).

For quadrupolar nuclei, with spin $I > 1/2$ (such as ^{17}O), the relaxation is caused by the interactions of the nuclear quadrupole moment with the surrounding fluctuating electrical field. In the case of deuterium (^2H) NMR, however, chemical exchange also contributes to the nuclear spin relaxation.

Nuclear spin relaxation is therefore dependent upon the *molecular dynamics* of the liquid, which is characterized by a correlation time, or distribution of correlation times: the faster the motions are, the shorter the correlation time and the longer the relaxation time. The motion of a molecule, such as rotation around a specific axis, or a shift in a specific direction, occurs in a determined interval of time that is the correlation time of that particular motion. The random isotropic reorientation of a molecule is, however, the type of motion usually considered in the nuclear spin relaxation studies of liquids. In this case, a single average correlation time for the isotropic reorientation, τ_c, provides the simplest estimate of the *molecular mobility* in the liquid.

Basic Deuterium NMR Theory

The discussion of the results presented in the following sections requires a brief look at the basic ^2H NMR theory, with emphasis on the ^2H NMR spectra of solids.

An isolated deuteron with a nuclear spin $I = 1$ in a single-crystal rigid lattice would have three equally spaced, Zeeman energy levels, *if and only if*, the electrical quadrupole interaction was neglected in the calculation (Fig. 5-18a). The presence of an appreciable quadrupole interaction (which is the real case for any deuterium, with only one exception), however, requires the calculation of (Zeeman + quadrupole) deuteron spin energy levels with a *quadrupolar Hamiltonian spin operator* (according to Abragam, 1961):

$$H_Q = \frac{e^2qQ}{4I(2I - 1) \cdot h} \cdot \frac{3\cos^2\theta - 1}{2} \cdot [3m^2 - I(I + 1)] \qquad (5\text{-}6)$$

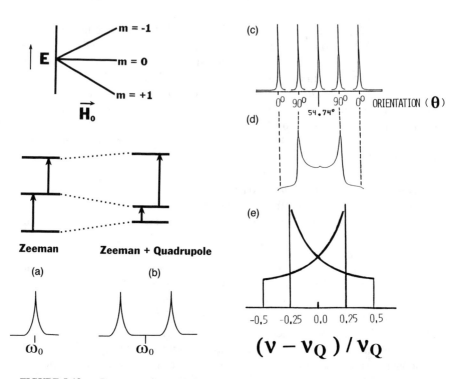

FIGURE 5-18. Deuteron spin energy levels in a static magnetic field, in the absence (a) and presence (b) of quadrupolar interactions. (c) The orientation dependence of ^{2}H NMR spectra of a deuteron in a single crystal. (d) The ^{2}H NMR spectrum of a deuteron in a polycrystalline powder. (e) Calculated ^{2}H NMR powder spectrum, neglecting dipolar interactions.

where $I = 1$; (e^2qQ/h) is the quadrupole coupling constant; (eq) is the electric field gradient at the deuteron; (eQ) is the deuteron quadrupole moment; $m = 1, 0,$ or -1 is the magnetic quantum number for the deuterium; θ is the angle that defines the orientation of the single crystal with respect to the static magnetic field; and h is Planck's constant. The single-deuteron spin energy levels obtained with the H_Q operator given in equation (5-6) are presented in Figure 5-18b. Notably, the presence of the electrical quadrupole interaction causes a rise in the $m = +1$ and $m = -1$ spin energy levels, whereas it lowers the $m = 0$ energy level. The separation of the spin energy levels is thus *unequal* and is *orientation dependent* (Fig. 5-18a to d). There are therefore only two possible single-quantum transitions ($\Delta m = \pm 1$) for the isolated deuteron (that is, a deuteron that is *not* interacting with any deuterons or other nuclei), as shown in the NMR spectrum of the single deuteron at the right in Figure 5-18b. The angular, or orientation dependent, $(3\cos^2\theta - 1)/2$ term in equation (5-6) will cause the doublet of the deuteron to change with the orientation (θ) of the $\mathbf{X} \rightarrow \mathbf{D}$ bond vector in the single crystal with respect to the external magnetic field vector $\vec{\mathbf{H}}_0$, in the manner shown in Figure 5-18c. Note that at the "magic angle," (θ_0 = 54.74°), the angular term $(3 \cos^2 \theta_0 - 1)/2$ vanishes, and there is only a single peak located in the center, instead of a doublet.

In the case of deuterium NMR of crystalline powders, the spectrum is obtained by averaging the angular term over all possible orientations. In the resulting powder pattern, the 90° orientation dominates, whereas the 0° peaks appear as edges, or singularities, symmetrically placed with respect to the center of the powder pattern, and with twice the spacing of the 90° splitting, as shown in Figure 5-18d. As an example, the two deuterons of a water molecule in a D_2O (hexagonal structure, I_h) ice powder give a spectrum like the one shown in Figure 5-18d, with a splitting $\Delta \nu_Q$ between the two central peaks of ~160 kHz in the rigid lattice (at 100 K). At the other extreme, in liquid D_2O at room temperature (293 K), the quadrupolar splitting completely collapses, because the quadrupolar interaction is averaged by the extremely fast isotropic (random) motions of D_2O in the liquid, and only one single Lorentzian ^2H NMR peak of ~0.5 Hz half-height linewidth is observed for both deuterons in the D_2O molecule. Thus, the deuterium NMR relaxation in the liquid is essentially *quadrupolar*. The correlation time of liquid D_2O calculated from this single Lorentzian ^2H NMR peak is about 4.7 ps at 293 K (20°C).

Between these two extremes of a rigid solid and a liquid there are many intermediate cases possible, two of which are especially significant to the analysis of the ^2H NMR data presented in the next section. The fast rotation of the D_2O molecule about its symmetry axis would substantially reduce the quadrupole splitting from the value measured in the rigid ice lattice, which would also cause other changes in the lineshape (Fig. 5-19). Static disorder, caused by variations of the $O-D$ bond orientations throughout the lattice of a solid, would also cause a reduction in the deuteron quadrupole splitting for the D_2O molecule. Such variations in molecular orientations of the $\mathbf{X} \rightarrow \mathbf{D}$ bond vector can be characterized by one- to three-order parameters, depending on the types of misorientation encountered. For D_2O molecules in a lattice the variation in the orientation of the D_2O molecule symmetry axis throughout the lattice can be represented by a single-order parameter, S. The reduction in the quadrupolar splitting caused by such static disorder can be calculated from a powder spectrum with the following equation:

$$\Delta \nu_q = \left(\frac{3}{4} \right) \cdot \frac{e^2 qQ}{h} \cdot S \cdot \tau_c \tag{5-7}$$

where $\Delta \nu_q$ is the residual quadrupole splitting and τ_c is the average correlation time of the D_2O molecules in the partially disordered lattice. For a completely

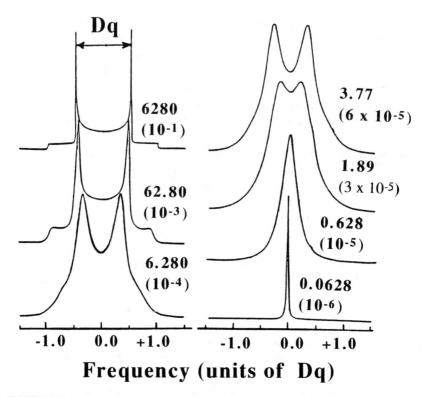

Frequency (units of Dq)

FIGURE 5-19.   Deuterium NMR spectra computed as a function of the order parameter and molecular motion (according to Smith, 1983).

random, amorphous solid, $S = 0$, whereas for a perfect crystal lattice, with no variation in the orientation of the D_2O molecules, $S = 1.0$. It is interesting that even for relatively small values of S, the deuterium NMR powder patterns retain sharp features in the absence of fast motions (Fig. 5-19 and Smith, 1983).

Deuterium NMR lineshapes are also markedly affected by exchange processes, such as the rotational jumps, or ''flips,'' of deuterons among alternate sites. An example is shown in Figure 5-20A, where the ^2H NMR lineshapes are presented as a function of the exchange rate (or flip rate) of deuterons in an aromatic ring system (Smith, 1983) ($e^2qQ/h = 180$ kHz and $\eta = 0.06$). Large-angle flips yield unique ''double horn'' features (Fig. 5-20B) when these flips occur about an axis that is itself moving. Such a dynamic model is intuitively appealing for aromatic rings trapped within channel clathrates (or embedded in highly ordered smectic mesophases) and for phenyl group substituents in proteins. This type of model can be generalized to include more orientations around the flip axis to simulate the motions of adsorbates trapped near surfaces or at catalytically active sites.

EXCHANGE RATE

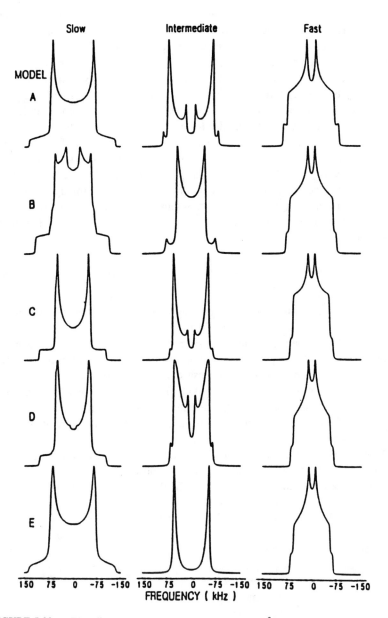

FIGURE 5-20. The effects of exchange rate on the computed ^{2}H NMR spectra (according to Jaffe, Vold, and Vold, 1982).

Electron Spin Resonance Spectroscopy

The electron spin resonance (ESR) technique (also called electron paramagnetic resonance (EPR)) is concerned with the resonant absorption of microwave energy by unpaired electrons in a material, corresponding to the quantum-allowed transitions between the unpaired electron spin ($S = 1/2$, $m_s = \pm 1/2$) energy levels in the presence of an external magnetic field, $\vec{H}_0$. Such transitions are induced by the absorption of resonant microwaves (of the "correct" frequency for resonance to occur). The microwave has an oscillating magnetic field component that is oriented perpendicular to the external magnetic field direction in order to be able to induce the ESR transitions. In practice, an ESR spectrometer operates with a klystron source of continuous microwaves (c.w. mode), set at a fixed frequency, ν_0, that irradiates through a waveguide that is placed in a specially designed ESR cavity between the poles of an electromagnet. The resonance condition is met by varying the magnetic field at a constant, slow rate until the splitting between the electron spin energy levels matches the microwave energy $\epsilon = h\nu_0$, so that

$$h\nu_0 = \gamma_e H_0 = g\beta_e H_0 \cdot S \tag{5-8}$$

where h is Planck's constant, γ_e is the magnetogyric ratio that has the value of 1.7×10^7 rad $\cdot$ G^{-1} s^{-1} for the free electron, g is a constant for the free electron equal to 2.000232, and β_e is the Bohr magneton of the electron. The resonance frequencies in ESR are typically of the order of 10 GHz, which is about 100 to 1000 times higher than in NMR because $\beta_e/\beta_n \simeq 1000$. Since the ESR measurements are carried out in the c.w. mode, detection is achieved by modulating the signal to increase sensitivity; the detected signal is therefore recorded as the derivative of the ESR absorption signal, and is plotted as the derivative amplitude against the variable magnetic field.

An example of an EPR spectrum of a stable (di-tert-butyl nitroxide) radical is shown in Figure 5-21a for a solution in ethanol; the corresponding transitions (Fig. 5-21c) are also indicated below this EPR spectrum. Because the unpaired electron spin is essentially localized around the nitroxyl group ($\rangle\dot{N}$—O) nitrogen, the effective magnetic field at the unpaired electron spin is $H_{eff} = H_0 - a_0 m_I$, where $m_I = -I$, $(-I + 1)$, . . . , $(I - 1)$, $(\pm I)$ and I is the nuclear spin quantum number of nitrogen (^{14}N, in this case $I = 1$). There are, in general, $(2I + 1)$ lines (Fig. 5-21c) in the corresponding EPR spectrum (Fig. 5-21a, b), and a_0 is a constant chosen so that an I represents the magnitude of the local field arising from the magnetic moments of the nuclei (^{14}N) in close proximity to the unpaired electron spin of the $\rangle\dot{N}$—O group. The parameter a represents the separation in gauss (G) between the EPR lines in the spectrum of Fig. 5-21a, and is called the *hyperfine splitting constant*. Wavefunctions of

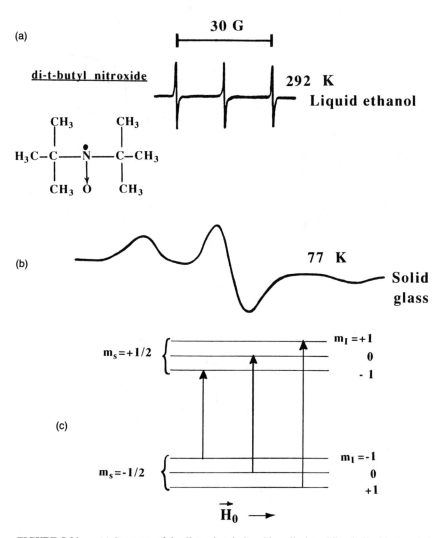

FIGURE 5-21. (*a*) Spectrum of the di-*tert*-butyl nitroxide radical tumbling in liquid ethanol at 292 K and (*b*) in a solid glass at 77 K (modified from Cantor and Shimmel, 1983). (*c*) The corresponding hyperfine splittings/electron spin transitions are indicated below the EPR spectrum. The m_I are the ^{14}N nuclear spin levels in the $\dot{\text{N}}$—O group.

electron orbitals of p, d, f (or higher) types have a node at the nucleus, and the electron distribution is strongly *orientation dependent* (see also Figs. 3-8 and 3-10), or anisotropic. In such cases, the corresponding hyperfine interaction is also anisotropic, and the *anisotropic couplings* along the principal axis system

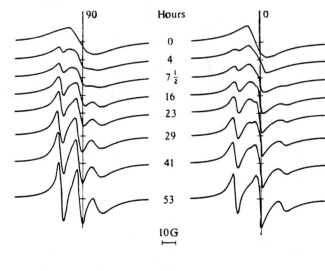

90 | Hours | 0

0

4

7 ½

16

23

29

41

53

10G

(a)

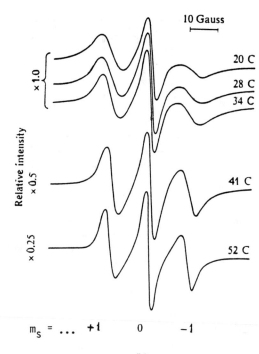

10 Gauss

20 C

28 C

34 C

× 1.0

41 C

× 0.5

Relative intensity

× 0.25

52 C

$m_s = \ldots \quad +1 \qquad 0 \qquad -1$

(b)

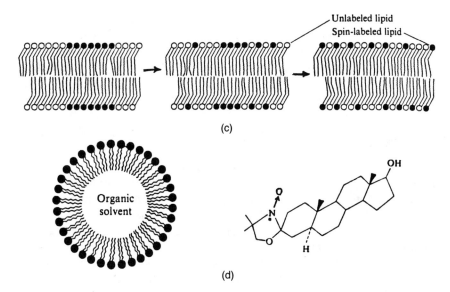

Unlabeled lipid
Spin-labeled lipid

(c)

Organic
solvent

OH

(d)

FIGURE 5-22. (*a*) Paramagnetic resonance spectra of spin-labeled vesicles at various times after the onset of lateral diffusion (from Devaux and McConnell, 1972). (*b*) EPR spectra of spin-labeled steroids in monolayer vesicles at various temperatures. Magnetic quantum numbers (+1, 0, −1) of the three lines are indicated (after Sackmann and Träuble, 1972). (*c*) Procedure for the measurement of lateral diffusion rates (after Devaux and McConnell, 1972). (*d*) Lipid monolayer vesicles—schematic diagram. (Also shown is the chemical structure of the spin-labeled steroid.)

(x, y, z) are specified as a_x, a_y, and a_z, respectively. The anisotropic couplings can be determined from EPR studies of oriented single crystals. The values of the anisotropic (hyperfine) couplings for the example considered in Figure 5-21*a*, *b* are $a_x = 7.1 + 0.5$ G, $a_y = 5.6 \pm 0.5$ G, and $a_z = 32.0 \pm 1.5$ G; the corresponding value of a_0 for the di-tert-butyl nitroxide radical tumbling rapidly in the ethanol solution is obtained by averaging the a_x, a_y, and a_z values:

$$a_0 = (1/3)(a_x + a_y + a_z) \qquad (5-9)$$

Because most materials do not possess electron paramagnetism (they do not have unpaired electron spins), they can only be studied after the inclusion or covalent attachment of selected spin labels. Stable nitroxyl radicals are very popular as spin labels for polymer and biopolymer studies by EPR. Phospholipid bilayers and biological membranes can also be studied in detail by employing spin-labeled hydrophobic probes that contain an attached nitroxyl group, which, in this case, would give an EPR spectrum. Two series of examples of EPR spectra for spin-labeled lipid vesicles (liposomes and lipid monolayer vesicles) are presented in Figure 5-22*a* and *b*, respectively. In Figure 5-22*a* one can see the time dependence of the (slow) *lateral diffusion* of the spin labels in the lipid bilayer, whereas in Figure 5-22b one can follow the temperature-induced changes in the motions of spin-labeled steroids within the lipid monolayer vesicles. Other interesting applications of this sensitive and powerful technique to polysaccharides of interest to the food technologist are discussed in greater detail in the second volume of this book (in Chap. 11).

THERMOANALYTICAL TECHNIQUES

The thermoanalytical techniques are a group of procedures in which the temperature dependence of a physical property of a material is measured by subjecting the material to a controlled temperature program. This group includes differential scanning calorimetry (DSC), differential thermal analysis (DTA), thermomechanical analysis (TMA), and thermogravimetric analysis (TGA). Both DSC and DTA are being utilized in food research, and, therefore, only these are discussed here in relation to food applications. Also, some of the thermodynamic concepts introduced in Chapter 2 of this volume are employed in these discussions.

Differential Scanning Calorimetry

Heat processing is often used to manufacture food products. Among major food applications that involve heat processing and heat effects are cooking, extrusion cooking, sterilization/pasteurization, and drying. Heat extraction is also employed in freezing, freeze-drying, and cold storage, to prolong the shelf life of food products. The effects of heat or heat extraction on the *functionality* of food materials are therefore of great interest to the food industry and food technologists. A basic understanding of the effects of heat exchange and related thermodynamic processes (phase transitions, etc.) on foods is necessary in order to reduce production costs and increase profits. Thermoanalytical techniques, such as differential scanning calorimetry (DSC), provide a wealth of information that needs to be correctly interpreted in thermodynamic terms.

In both DSC and differential thermal analysis (DTA) measurements a sample and a reference material are simultaneously measured to determine the effects of temperature while both are subjected to a controlled temperature program (Fig. 5-23a and b, respectively). In the case of DTA, one measures the temperature difference in such a process, whereas in DSC, one measures the difference between the energy inputs to the sample and the reference. Two modes of measurement are common in DSC: a power-compensation method and a heat-flux method. In the former method, one measures the compensatory power flow, or the energy required to maintain a temperature (T) in null-balance between the sample and the reference by compensating for heat exchange. In prac-

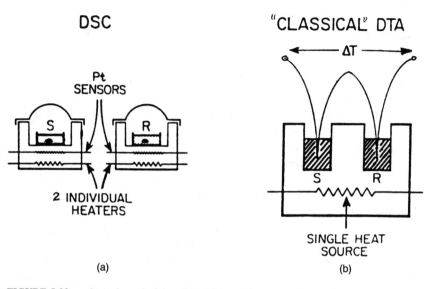

FIGURE 5-23.   Operating principles of (*a*) DSC and (*b*) DTA instruments, respectively (courtesy of Perkin-Elmer Corporation).

tice, DTA employs a single heat source, whereas DSC employs two separate heat sources, one for the sample and one for the reference, as well as two platinum sensors to measure temperature (Fig. 5-23*a*).

The output of a DSC (or DTA) instrument is a thermogram, which is a plot of the differential heat flow (dQ/dT) against temperature. A change in the dependent variable dQ/dT in fact represents an absorption, or evolution, of heat by the sample, that is, an *enthalpy change* (for a definition of enthalpy, see Chap. 2):

$$\frac{dQ}{dt} = \frac{dH}{dT}$$

where t is time. At constant pressure one can determine the heat capacity, C_P, as:

$$C_p = \frac{dH}{dT} = \frac{dQ}{dT} \times \frac{dt}{dT} \tag{5-10}$$

where $(dt/dT)^{-1}$ is the programmed heating rate. (For a more detailed discussion of heat capacity, see Chap. 2.)

By weighing the sample before and after the measurement, one can express the dependent variable in terms of the specific heat $C_{ps} = C_p/m$, if the sample mass is *m* and does not change during the measurement. In practice, one measures the specific heat of a food material relative to that of a known sapphire standard (or reference). This specific heat determination by DSC is represented schematically in Figure 5-24.

Whenever *phase transitions* or *structural changes* occur in a material due to changes in the temperature, the *heat capacity* of the material also changes, because the material heat-exchange properties depend upon structure and component distribution in the material. For example, the denaturation of proteins by heat results in a sharp change in the heat capacity of the protein, while prior to denaturation heat capacity increases gradually. The latter process was interpreted to be the result of the gradual unfolding of the protein, resulting in an increased number of side chains being exposed to water (Privalov et al., 1971). In food research, the observed DSC transitions may relate to processes such as protein denaturation, gelling, starch gelatinization, fat crystal melting, or other phase transitions. Relatively sharp DSC peaks are observed for cooperative processes such as protein denaturation. A simplified example is drawn schematically in Figure 5-25, where, after appropriate corrections, the integrated peak area represents the enthalpy change $\Delta H = \int_{T_1}^{T_2} \Delta C_p \cdot dT$ if the ordinate is cali-

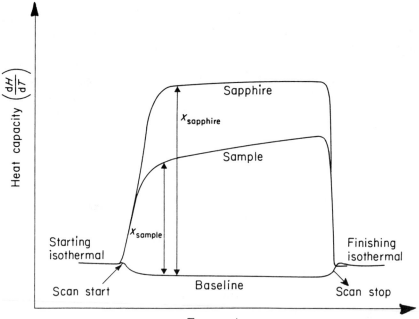

FIGURE 5-24.   Determination of specific heat by DSC. A schematic representation of specific heat determination is shown in Figure 5-25 (from Wright, 1984, with permission).

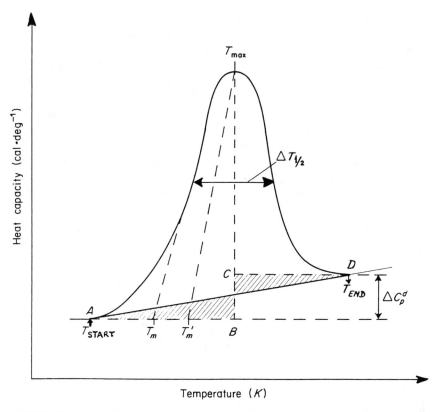

FIGURE 5-25.   A DSC thermogram of a transition. The derivation of the parameters from the DSC peak is also schematically indicated (modified from Wright, 1984).

brated in units of heat capacity. The temperature at the peak, T_{max}, is also of interest as a *"transition temperature,"* but it does depend on the rate at which the temperature is being changed (slope A–D in Figure 5-25). For a two-state process one can estimate the *Van't Hoff enthalpy*, $\Delta H_{VH} = (4RT_M^2/\Delta T_{1/2}$, by measuring the transition midpoint temperature and the peak width at half height. The value of ΔH_{VH} can then be compared with ΔH_{cal}, the calorimetrically determined enthalpy, in an adiabatic or isothermal calorimeter to indicate whether or not the transition really corresponds to a two-state process.

Kinetic parameters and relatively slow reaction kinetics can also be estimated by DSC; however, they are not discussed in this volume.

DSC was employed to distinguish between the thermal stability of proteins and their tendency to renature after cooling. Proteins present in milk whey, such as α-lactalbumin, xanthine oxidase, ribonuclease, and lysozyme, were shown by DSC to denature at relatively low temperatures, but rescanning indicated that they slowly renatured after cooling to room temperature. Hydration numbers of proteins of about 32% to 34% weight-to-weight ratio (w/w) or more were also suggested by DSC measurements.

Precise results and sharp transition peaks were observed by DSC for a variety of lipids in their hydrated state. The tall endothermic peak for hydrated phospholipids was assigned to the gel-to-liquid-crystalline phase transition involving cooperative "melting" of the hydrocarbon chains of phospholipid molecules (Ladbroke and Chapman, 1969). This transition is accompanied by a rather sharp DSC peak (Fig. 5-26), indicating the high degree of *cooperativity* present in such systems as a result of hydrophobic interactions. The enthalpy change of the transition increased along with the transition temperature for phosphatidycholines of increasing acyl chain lengths from 12 to 22 carbon atoms (C_{12} to C_{22}) (Mabrey and Sturtevant, 1978). In foods, the polymorphism of commercial fats and oils is often studied by DSC. Because of the complexity of

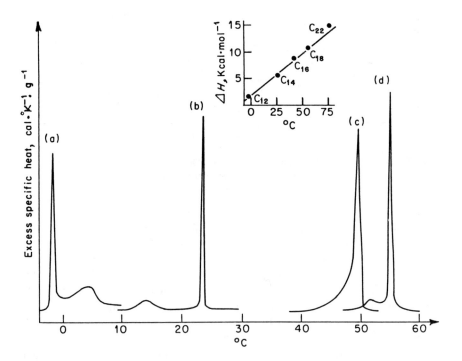

FIGURE 5-26.   DSC thermograms of (*a*) dilauroylphosphatidylcholine, (*b*) dimyristoylphosphatidylcholine, (*c*) dymyristoylphosphatidylethanolamine, and (*d*) distearoylphosphatidylcholine. The inset shows the relationship between temperature and the enthalpy of transition for phosphatidylcholines with acyl groups containing from 12 to 22 carbon atoms. The heating rate in the experiment was $1°$ min^{-1}. (Redrawn from Mabrey and Sturtevant, 1978, with permission; copyright © 1978 Plenum Publishing Corp., New York.)

DSC analysis of polymorphic lipid or fat-containing foods, the DSC measurements are often paralleled by X-ray diffraction in programmed temperature studies (Chapman et al., 1971). The identification of various oils in mixtures and the differentiation of lard from tallows can also be carried out by DSC, which can also be used to determine the solids content in fats and the free fatty acid content in several foods. These are only some of the applications of the technique to food. Somewhat more controversial are those DSC studies concerning starch gelatinization, where equivalent X-ray diffraction studies of the DSC samples could help substantiate the interpretation of the DSC/DTA thermograms. Nevertheless, it is rather interesting that certain higher-amylose starches tend to have higher gelatinization temperatures, and that the gelatinization endotherm only appears in potato starch when water is present in excess of ~ 32% (Duckworth, 1971). The threshold of 32% moisture content corresponds to *four molecules* of water per glucose unit, or to the amounts of *nonfreezable water* present.

Various problems with the correction and interpretation of DSC thermograms of foods were reviewed by Wright (1984).

References

Abragam, A. 1961. *The Principles of Nuclear Magnetism.* Oxford, England: Clarendon Press.

Baianu, I. C. 1974. Structural Studies of Erythrocyte and Bacterial Cytoplasmic Membranes by X-ray Diffraction and Electron Microscopy. Ph.D. thesis, University of London.

Baianu, I. C., et al. 1992. Molecular dynamics, local structure and spectroscopy of liquid water, glassy water and the liquid water-vapor interface. In *Interfacial Water in Dispersed Systems*, W. Drost-Hansen, ed. Berlin and New York: Springer-Verlag.

Brey, W. S. 1972. *Physical Methods for Determining Molecular Geometry*, 3d ed. New York: Van Nostrand Reinhold.

Cantor, C. R., and H. Shimmel. 1982. *Biophysical Chemistry*, 3 vols. San Francisco: W. H. Freeman.

Chapman, G. M., E. E. Ackhurst, and W. B. Wright. 1971. Cocoa butter and confectionery fats. Studies using programmed temperature X-ray diffraction and differential scanning colorimetry. *J. Am. Oil Chem. Soc.* **48:**824–830.

Crick, F. H. C. 1952. Is α-keratin a coiled-coil? *Nature* **170:**882–883.

Crick, F. H. C. 1953. Fourier transform of a coiled-coil. *Acta Cryst.* **6:**685–687.

Crick, F. H. C., and J. D. Watson. 1956. Structure of small viruses. *Nature* **177:**473–476.

Devaux, P., and H. M. McConnell. 1972. Lateral diffusion in spin-labeled phosphatidylcholine multilayers. *J. Am. Chem. Soc.* **94:**4475–4481.

Duckworth, R. B. 1971. Differential thermal analysis of frozen food systems. The determination of unfreezable water. *J. Food Technol.* **6:**317–327.

Finean, J. B. 1967. *Engström-Finean Biological Ultrastructure*, 2d ed. London and New York: Academic Press.

Franklin, R. E. 1956. Location of the ribonucleic acid in the tobacco mosaic virus. *Nature* **177**:928–930.

Franklin, R., and R. G. Gosling. 1953. Evidence for 2-chain helix in crystalline structure of sodium deoxyribonucleate. *Nature* **171**:156–157.

Jaffe, D., R. L. Vold, and R. R. Vold. 1982. Deuterium relaxation of a partially oriented methyl group. *J. Magnetic Resonance* **46**:475–496, 496–504.

Jarell, H. C., and I. C. P. Smith. 1983. Deuterium NMR spectroscopy. In *The Multinuclear Approach to NMR Spectroscopy*, J. B. Lambert and F. G. Riddell, eds., NATO ASI series C103. Dordrecht, Germany: D. Reidel.

Ladbroke, B. D., and D. Chapman. 1969. Thermal analysis of lipids, proteins and biological membranes. A review and summary of some recent studies. *Chem. Phys. Lipids* **3**:304–367.

Levine, Y. K., and M. H. F. Wilkins. 1971. Structure of oriented lipid bilayers. *Nature New Biology* **230**:69–72.

Mabry, S., and J. M. Sturtevant. 1978. High-sensitivity differential scanning calorimetry in the study of biomembranes and related model systems. *Methods Membrane Biol.* **9**:237–274.

Privalov, P. L., N. N. Khechinashvili, and B. P. Atanasov. 1971. Thermodynamic analysis of thermal transitions in globular proteins. I. Calorimetric study of chymotrypsinogen, ribonuclease and myoglobin. *Biopolymers* **10**:1865–1890.

Sackmann, E., and H. Träuble. 1972. Studies of the crystalline-liquid crystalline phase transition of lipid model membranes. I: Use of spin labels and optical probes as indicators of the phase transition; II: Analysis of electron spin resonance spectra of steroid labels incorporated into lipid membranes; III: Structure of a steroid-lecithin system below and above the lipid-phase transition. *J. Am. Chem. Soc.* **94**:4482–4510.

Smith, I. C. P. 1983. Deuterium NMR. In *NMR of Newly Accessible Nuclei*, P. Laszlo, ed., 1–32. New York: Academic Press.

Soper, K., and M. G. Phillips. 1986. A new determination of the structure of water at 25°C. *Chem. Phys.* **107**:47–60.

Vold, R. L., J. S. Waugh, M. P. Klein, and D. E. Phelps. 1968. Measurements of spin relaxation in complex systems. *J. Chem. Phys.* **48**:3831–3832.

Watson, J. D., and F. H. C. Crick. 1953a. Genetical implications of the structure of deoxyribonucleic acid. *Nature* **171**:964–967.

Watson, J. D., and F. H. C. Crick. 1953b. Molecular structure of nucleic acids: A structure for deoxypentose nucleic acids. *Nature* **171**:737–738.

Wilkins, M. H. F., A. R. Stokes, and H. R. Wilson. 1953. Molecular structure of nucleic acids: Molecular structure of deoxypentose nucleic acids. *Nature* **171**:738–739.

Wilkins, M. H. F., W. E. Seeds, A. R. Stokes, and H. R. Wilson. 1953. Helical structure of crystalline deoxypentose nucleic acid. *Nature* **172**:759–762.

Wilson, H. R. 1966. *Diffraction of X-rays by Proteins, Nucleic Acids and Viruses*. London: Arnold.

Wright, D. J. 1984. Thermoanalytical methods in food research. In *Biophysical Methods in Food Research*, H. W.-S. Chan, ed., 1–36. Oxford, England: Blackwell.

Zobel, H. F. 1988. Molecules to granules: A comprehensive starch review. *Starch* **40**:44–65.

II

Applications to Food Processes

6

Molecular Interactions and Processes in Foods

The functionality of foods depends just as much on the various *interactions* among food components as it does on their *structure*. Processes occurring in foods are essentially physical or chemical/biochemical ones. Chemical reactions in foods have been and still are intensely studied; consequently, there is a large body of knowledge regarding the complex chemical reactions that occur in foods, although it is by no means complete. The emphasis in this chapter is on those processes in foods that do not result in major chemical changes. A few examples of free-energy change calculations for chemical/biochemical reactions are also presented.

Molecules in foods interact through *electrostatic forces, hydrogen bonding*, and van der Waals and *hydrophobic interactions*. The energy of interaction is the largest for electrostatic forces and the smallest for hydrophobic effects, as indicated in Table 6-1. A quantitative treatment of all these interactions in complex food systems is an extremely difficult, if not impossible, task. However, some of these interactions can be treated numerically with reasonable success, especially in the simpler systems. During the last ten years, *molecular dynamics* (MD) computations on a fast computer/supercomputer have become increasingly successful in approaching more complex systems, local structure and dynamics in liquids, phase transitions, and surface phenomena. Precise, analytical solutions are available only for very simple systems such as *dilute* electrolyte solutions (at concentrations less than about 10 to 50 mM). For the latter systems, the *Debye theory* provides satisfactory answers except for *specific* interactions between ions that it is unable to take into account. Such specific interactions among molecular components and ions are especially important in biochemical systems and foods. Numerous efforts have been made to develop quantitative approaches to *concentrated* electrolyte solutions and *condensed*

TABLE 6-1   Types of Noncovalent Forces Important for Protein Structures

Type	Example		Binding Energy (kcal/mol)	Change of Free-Energy Water→Ethanol (kcal/mol)
Dispersion forces	Aliphatic hydrogen	$-\overset{\mid}{\underset{\mid}{C}}-H \cdots H-\overset{\mid}{\underset{\mid}{C}}-$	-0.03	
Electrostatic interaction	Salt bridge	$-COO^{\ominus} \cdots H_3\overset{\oplus}{N}-$	-5	-1
	Two dipoles	$C^{\delta+}=O^{\delta-} \cdots O^{\delta-}=C^{\delta+}$	$+0.3$	
Hydrogen bond	Ice	$O-H \cdots O$	-4	
	Protein backbone	$N-H \cdots O=$	-3	
Hydrophobic forces	Side chain of Phe			-2.4

Source: Reproduced with permission from Schultz and Schirmer, 1979.

systems with more than two components. Efforts are also being made to quantitate the effects of *hydrophobic interactions* and *nonideal* effects in electrolyte solutions.

Quantitative treatments of polyelectrolyte solutions and nonideal effects were successfully introduced by Kirkwood and Shumaker (1952a; 1952b), and subsequently by Tanford (1961), and Timasheff and Coleman (1960). This approach is especially relevant to food systems that often include proteins and ions in relatively high concentrations. The application of the Kirkwood and Shumaker theory to the analysis of nuclear spin relaxation data for certain food proteins in aqueous solutions was introduced by Kumosinski and Pessen (1982).

Although electrostatic interactions are the strongest, there are several systems in which hydrophobic interactions and hydrogen bonding dominate the system properties. (Note, however, that hydrogen bonding involves electrical interactions between *polar*, rather than charged, groups; hence, the lower interaction energy involved in hydrogen bonding by comparison with electrostatic interactions.) Van der Waals interactions are considered first, followed by consideration of an example of hydrogen bonding in liquid water, hydrophobic interactions, and representative treatments of electrostatic interactions in systems such as electrolyte and polyelectrolyte solutions that are relevant to foods.

VAN DER WAALS INTERACTIONS

Dispersion Forces and Repulsion between Electron Orbitals

Any pair of two atoms are subject to attractive interactions when the atoms are close to each other. This action is caused by the fact that each atom behaves like a (transient) oscillating (electrical) dipole generated by the relative motion of the electrons with respect to the nucleus. Such an oscillating dipole of an atom induces a transient oscillating dipole of opposite orientation in its neighbor atom, a process called *polarization*. Because of the opposite orientation of these oscillating dipoles, the forces exerted on the atom pair are *attractive;* such attractive interactions are called *dispersion forces.* The attractive energy of dispersive forces decreases very rapidly with increasing distance between the atoms and, to a first approximation, it is proportional to the sixth power of the inverse distance between the atoms, as well as the polarizability of the atoms. Dispersion forces vary, in general, with the relative orientation of molecular groups, because the atoms bound in a molecule have an anisotropic polarizability related to the orientations of the molecular orbitals. Because the orientation effects are tiny, however, they are usually neglected and the dispersion forces are treated as being isotropic. An example of a *dispersion force* is that between two aliphatic groups

$$
\begin{array}{ccc}
| & & | \\
-\text{C}-\text{H} & \cdots & \text{H}-\text{C}- \\
| & & |
\end{array}
$$

which has a corresponding binding energy of about -0.03 kcal/mol.

Electron orbitals are also subject to strong electrostatic repulsion forces that have an appreciable value at very short distances, and decrease rapidly with increasing distance between the electron shells. The corresponding Lennard-Jones term often is taken to be proportional to the twelfth power of the inverse distance between the atoms. When combined with London's $(1/R^6)$ term for the dispersion forces, the Lennard-Jones $(1/R^{12})$ term leads to a relatively simple equation for the interaction energy between the electron orbitals of neighboring atoms:

$$ E = \frac{A}{R^{12}} - \frac{B}{R^6} = E_m\left[-\left(\frac{R_m}{R}\right)^{12} + 2\left(\frac{R_m}{R}\right)^6 \right] \qquad (6\text{-}1) $$

where $E_m = -B^2/4A$ and $R_m = \sqrt[6]{(2A/B)}$ correspond to the minimum (equilibrium position) value of this function. The preceding Lennard-Jones/London

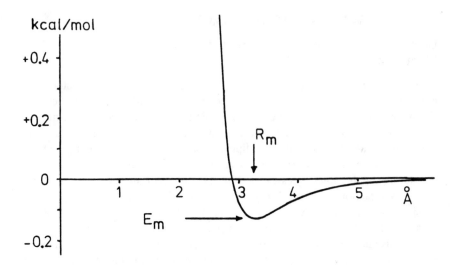

FIGURE 6-1.   Lennard-Jones 6-12 potential for dispersion forces and electron repulsion (calculated with $R_m = 3.24$ Å and $E_m = -0.13$ kcal/mol). Because of the repulsion between electronic orbitals and attraction via dispersion forces; in this case, the parameters A and B are both positive, and there is a relative minimum present in this potential (from Schulz and Schirmer, 1979, with permission).

(LJ/L6-12), 6-12 interaction energy is illustrated in Figure 6-1 as a function of the internuclear distance, R. Note the marked asymmetry of the potential well. Calculated values of the LJ/L6-12 parameters E_m and R_m are given in Table 6-2 for electron shell dispersion and repulsion forces at nonbonded contacts, according to Momany et al. (1975). This table shows that the energy contribution

TABLE 6-2   **Parameters of Lennard-Jones 6-12 Potential for Electron Shell Repulsion and Dispersion Forces at a Nonbonded Contact***

	Momany et al.		Lifson and Warshel	
	E_m (kcal/mol)	R_m (Å)	E_m (kcal/mol)	R_m (Å)
Aliphatic H . . . aliphatic H	−0.04	2.92	−0.01	2.94
Aliphatic C . . . aliphatic C	−0.04	4.12	−0.19	4.23
Carbonyl O . . . carbonyl O	−0.20	3.12	−0.23	3.00
Amide N . . . amide N	−0.11	3.51	−0.19	3.60

Source: Reproduced with permission from Schulz and Schirmer, 1979.

*Only contacts between identical types of atoms are considered. For a contact between nonidentical types i and j use $(E_m^{(ij)} = E_m^{(ii)} + E_m^{(jj)})^{1/2}$ and $R_m^{(ij)} = 1/2\,(R_m^{(ii)} + R_m^{(jj)})$

TABLE 6-3   Partial Charges of Atoms in the Polypeptide Backbone and in Three Side Chains*

Peptide	N	−0.36		Tyr	O_η	−0.33
Peptide	H_N	+0.18		Tyr	H_η	+0.17
Peptide	C_α	+0.06				
Peptide	C'	+0.45		Asn	C_β	−0.12
Peptide	O	−0.38		Asn	H_β	+0.06
Peptide	H_α	+0.02		Asn	C_γ	+0.46
				Asn	O_δ	−0.38
Ser	C_β	+0.13		Asn	N_δ	−0.45
Ser	H_β	+0.02		Asn	H_δ	+0.20
Ser	O_γ	−0.31				
Ser	H_γ	+0.17		Cys	S_γ	+0.01
				Cys	H_γ	+0.01

Source: Reproduced with permission from Schulz and Schirmer, 1979.

*Taken from the results of Momany et al., who derived the partial charges of the 20 common amino acid residues using the CNDO/2 (complete neglect of differential overlap) method. Partial charges were also derived from *ab initio* molecular orbital calculations of small molecules or by fitting observed crystal data. A comparison between the results shows that accuracy is low.

per atomic contact is quite small, but *not negligible*, especially for biopolymers where the *number of contacts* may be very large. A complete computation may also include "effective charges" q_i and q_j of the contact atoms i and j, which can be derived from partial charges (see, for example, the results of Momany et al. (1975) for proteins, presented in Table 6-3). Such effective charges add a repulsive, electrostatic term, $(q_i q_j / R)$, to the right side of equation (6-1). Calculated parameter values of the LJ/L6-12 van der Waals potential, derived from already crystallographically (X-ray) determined protein structures, are listed in Table 6-4. In the case of the amide nitrogens the electrostatic repulsion between the partial, negative charges of the nitrogens exceeds the attraction caused by the dispersion forces specified in Table 6-2.

TABLE 6-4   Parameters of a 6-12 van der Waals Potential as Derived from Observed Protein Structures

	A (kcal/mol · Å^{12})	B (kcal/mol · Å^6)	E_m (kcal/mol)	R_m (Å)
Aliphatic C . . . aliphatic C	2,750,000	+1425	−0.19	3.53
Carbonyl O . . . carbonyl O	417,000	+108	−0.01	3.96
Amide N . . . amide N	417,000	0	Repulsive	
Carbonyl O . . . carbon in benzene ring	695,000	−570	Repulsive	

Source: Reproduced with permission from Schulz and Schirmer, 1979.

TABLE 6-5   **Van der Waals Radii**
(as given by Bondi)

Type of Atom	Radius ($\mathring{A}$)
Aromatic H	1.0
Aliphatic H	1.2
O	1.5
N	1.6
C	1.7
S	1.8

Source: Reproduced with permission from Schulz and Schirmer, 1979.

Contact Distances and van der Waals Radii

Van der Waals *contact distances* can be defined on the basis of the van der Waals potentials previously discussed (eq. (6-1)). The experimentally determined van der Waals contact distances for atom pairs can be converted to "general van der Waals radii" of specific atoms if each contact distance is assumed to be the sum of two-atom-type specific radii. This is, of course, only a rather rough approximation, since such resulting "radii" are only averages over all kinds of contact partners. Examples of such van der Waals radii are given in Table 6-5.

HYDROGEN BONDS

A hydrogen bond occurs predominantly as a result of the electrostatic interactions between the partial, but large, positive charge of an atom in a group (such as the H-donor atom in an $O^{\delta-}-H^{\delta+}$ group) and the large negative, partial charge of a close neighbor (H-acceptor) atom (such as the O-atom in the $C^{\delta+}=O^{\delta-}$ group). These charges are usually the ends of electrical dipoles of molecular groups. The distance between the H-donor and its covalent neighbor, H-acceptor is significantly shorter than the sum of the expected van der Waals radii (Table 6-5). For example, the distance between an amide hydrogen and an H-bonded carbonyl oxygen is 1.9 Å instead of the 2.7 Å distance calculated as the sum of the van der Waals radii; hence, the entire electron shell of the hydrogen atom (which only has *one* electron) is shifted toward the atom to which the hydrogen is H-bonded (in this case the carbonyl oxygen). As a result, the shell repulsion between the H-bonded *contact* partners is quite small and the attracting charges approach each other more closely than would be expected from van der Waals considerations alone. Another example can be schematically represented as follows:

$$-O^{\delta-}-H^{\delta+}\cdots\cdot O^{\delta-}=C^{\delta+}$$

(F-6-1)

$$1\text{ Å}\quad 1.9\text{ Å}$$

where the positively charged (δ^+) H-atom is positioned between two negative, partially charged O-atoms, being closer to the oxygen on the left than to its H-bonded partner. The lowest potential energy for such a system of three charges occurs when all three charges are aligned. Therefore, hydrogen bonds tend to be *linear*. However, in liquids, there is enough thermal energy to bend some of the H-bonds; see, for example, liquid water, which is discussed next. A deviation of 20° in the H-bonding angle from the straight H-bond position reduces the binding energy by about 10%.

Only in special cases, such as $[F\cdots H-F]^-$, do the wavefunctions overlap, resulting in a "typical" covalent H-bonding, with a binding energy of 50 kcal/mol. The biologically important H-bonds, such as those in proteins, are much weaker, and there is no significant overlap of the wavefunctions of the H-bonded partners. Such H-bonds are *predominantly electrostatic*, as are those in liquid water and ice. Therefore, the H-bonds found in such systems tend to be relatively long (weak H-bonds), between 2.8 and 3.7 Å (Table 6-6), and are only 10% to 25% shorter than the sum of van der Waals radii for groups found in proteins. For liquid water the H-bond length is about 2.85 Å ($O\cdots O$ distance) compared with an expected van der Waals value of at least 3.12 Å; this length decreases to about 2.78 Å in ice I_h.

HYDROGEN BONDING IN LIQUID WATER

In a liquid the thermal motions cause the system to have relatively high entropy and relatively little ordering of the molecules. However, the close packing of the liquid molecules results in *local structures* that are not completely random. This is especially true when *directional* bonds exist in a liquid. In water, such directional bonds are provided by the hydrogen bonding between the neighbor water molecules. Since the water molecule has a tetrahedral arrangement of its sp^3 orbitals (see Chap. 4, the section on the geometry of the water molecule, and Fig. 4-11*a* and *b*), the local structure in liquid water is also quasi-tetrahedral. Water hydrogen bonds involve a one-pair orbital of one water molecule and a hydrogen atom of a neighbor water molecule. In ice I_h each water mol-

TABLE 6-6   Hydrogen Bonds Found in Proteins

Type of Hydrogen Bond		Distance between Donor and Acceptor Atom (Å)	Reduction from the Distance between Donor and Acceptor as Calculated from van der Waals Radii (%)	Comment
Hydroxyl–hydroxyl	$-$O$-$H$\cdots$O$_H$	2.8 ± 0.1	25	Bifurcated bonds are possible, for example, in ice. For phenols (Tyr), the length is 2.7 ± 0.1 Å
Hydroxyl–carbonyl	$-$O$-$H$\cdots$O=C$\backslash$	2.8 ± 0.1	25	
Amide–carbonyl	$>$N$-$H$\cdots$O=C$\backslash$	2.9 ± 0.1	20	Ubiquitous between main chain atoms in proteins; empirical length found in parvalbumin is 2.9 ± 0.3 Å
Amide–hydroxyl	$>$N$-$N$\cdots$O^H	2.9 ± 0.1	20	
Amide–imidazole nitrogen	$>$N$-$H$\cdots$N$\backslash$	3.1 ± 0.2	15	
Amide–sulfur	$>$N$-$H$\cdots$S$\backslash$	3.7	10	Found with Met-180 of chymotrypsin and at the FeS clusters of ferredoxin, rubredoxin, with lengths of 3.6 ± 0.3 Å

Source: According to Schulz and Schirmer, 1979.

ecule is hydrogen bonded to exactly *four neighbor* water molecules in a *tetra-hedral* arrangement around each water molecule. The hydrogen bonding in water causes the remarkable properties of liquid water and ice (highest density at $4\,°C$, high dielectric constant, relatively high melting and boiling points, etc.). The *local structure* of a liquid can only be characterized in a *statistical* manner by defining the probability, $g(r)$, that the first, second, third, and so on, nearest neighbors are found at given distances, r, from the water molecule chosen as a reference. This is, in fact, a *positional correlation function* that can be derived by Fourier transformation of the X-ray, or neutron, scattering intensity from a liquid sample (see also Chap. 5, the section on X-ray diffraction, and Vol. II of this book for further details regarding the scattering techniques). Narten, Danford, and Levy (1967) determined the X-ray *correlation function* of water, $G(r) = \Sigma_i\, g_i(r)$, where $g_i(r)$ are the *pair correlation* functions for the oxygen-oxygen, oxygen–hydrogen and hydrogen–hydrogen *atom pairs* in the liquid water. Because of the low scattering structure factor of the hydrogen, only the oxygen–oxygen pair correlation dominates the correlation function $G(r)$ of water. The X-ray scattering intensity functions of liquid water are shown in Figure 6-2*a* and a molecular model of the local structure in liquid water is shown in Figure 6-2*b*. Somewhat similar results were obtained by neutron scattering observations on heavy water (D_2O), as shown in Figure 6-2*c*.

HYDROPHOBIC INTERACTIONS

Nonpolar groups, or molecules, interact less with water molecules (and other polar solvents) than do water molecules among themselves, leading to what are called *hydrophobic interactions* between the nonpolar groups.

Hydrophobic interactions are important in micelle formation, lipid aggregation and structuring, in soaps ("colloidal" electrolytes), in food emulsions, and in stabilizing protein conformations. Thus, two basic situations occur.

First, at *infinite dilution* there is the interaction between a nonpolar group and water, or so-called *negative hydration*. Here, the spontaneous attraction between the functional group surrounding an ionic center and water is so small that the overall solubility is caused only by *long-range* charge-solvent polarization effects. The effect of the charge center on the solubility of the molecule can be understood if we compare, for example, the solubility of neopentane $C(CH_3)_4$ with that of a salt of $N^+(CH_3)_4$. Neopentane is insoluble in water, while the tetramethyl ammonium salts are quite soluble. In both cases, the methyl (CH_3) groups are, of course, hydrophobic (nonpolar). This is the *type I* situation.

Second, at *finite concentrations* of the nonpolar molecules (or "hydrophobic solute"), association of the hydrophobic molecules does occur in the presence

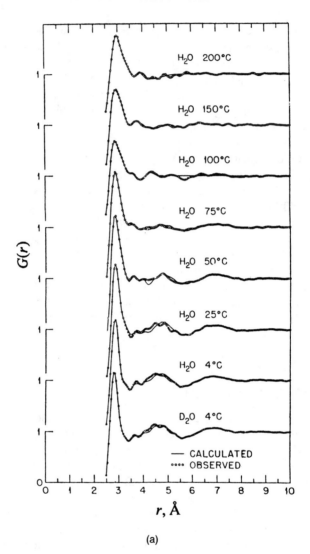

(a)

FIGURE 6-2.   (a) Radial distribution function (r.d.f.) of liquid water (H$_2$O) at various temperatures derived from X-ray scattering studies by Narten, Danford, and Levy (1967). (b) Representation of the local structure and hydrogen bonding in liquid water based on the r.d.f, $G(r)$, in part (a). This structural model is essentially an ice I_h model with interstitial water molecules being present to account for the increased density in liquid water in comparison with ice I_h at 0°C, as well as corresponding to the additional peak at about 3.2 Å in the r.d.f of liquid water in part (a) (from Narten, Danford, and Levy, 1967, with permission). (c) Calculated and experimentally determined pair distribution function $G_{O-O}(r)$ for the oxygen atoms in the liquid water. (Results based on the neutron scattering experiments by Soper and Phillips [from Soper and Phillips, 1986, with permission].)

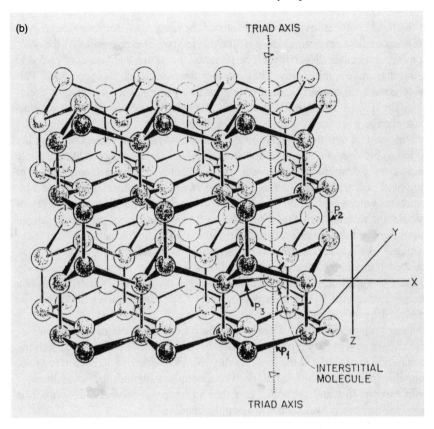

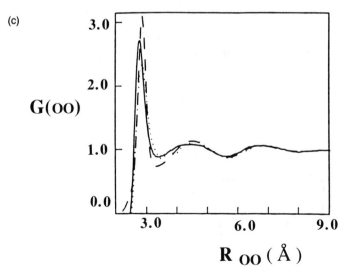

of the liquid solvent, even between ions of the same sign, for example, R_4N^+! This association determines most of the thermodynamic properties of the solutions, or molecular dispersions. The particles, or molecules, are not bonded to each other in a chemical way, although they are associated, or aggregated. This association is improperly called *hydrophobic bonding*, and is a kinetic event. Since the degree of order of the associated molecules increases, it has a significant change in *entropy*. This is the *type II* situation.

One view is that the association is a passive one, in the sense that there are no forces between the nonpolar molecules, and the association would occur only because two or more solute molecules or particles achieve a configuration of minimum potential energy by avoiding the energetically unfavorable type I interactions with water. In this model of hydrophobic interactions, the energy of association results from the minimization of the number of broken hydrogen bonds between the water molecules surrounding the hydrophobic molecule, or particle. This minimization results in the formation of a "cavity" for the associated particles rather than a cavity for each particle. However, in addition to the energy minimization effect, there is also the van der Waals interaction energy between the hydrophobic molecules, or particles, which is, in effect, the energy of *hydrophobic association*.

Type I interactions were detected by observing experimentally that the molar heat capacity of hydrophobic dispersions behaved anomalously, along with the partial molar entropy and the ionic fluidity changes occurring in certain organic, nonpolar (nonelectrolyte) solutions. For example, although the solubilities of hydrocarbons in water are much less than in nonpolar solvents, dissolution is more exothermic in water than in nonpolar media.

RADIAL DISTRIBUTION FUNCTION OF HYDROPHOBIC DISPERSIONS AND LIPID ASSOCIATION

Consider the example of $(n\text{-Bu})_4N^+F^-$ in H_2O. The local structure for the 'solution' $(n\text{-Bu})_4N^+F^-/H_2O$, derived from X-ray scattering studies by Narten, Danford, and Levy (1967), is *not* significantly *different* from that of liquid water alone! At present, there is no direct structural evidence of ordering of water caused by $(n\text{-Bu})_4N^+F^-$ in solutions. However, for aggregated lipids, there is clear X-ray and neutron scattering/diffraction evidence for the formation of lipid bilayers. Lipid bilayers are the basic structure in most biological membranes, including myelin, red-cell membranes, and bacterial membranes (see also Chap. 17 in Vol. II of this book). Hydrophobically induced changes in the local structure of water are nevertheless, as yet, unobserved.

MICELLAR OR "COLLOIDAL" ELECTROLYTES: ELECTRICAL PROPERTIES OF COLLOIDS

Aggregation of long-carbon-chain acids, or bases, through hydrophobic inter-actions leads to formation of the so-called *micelles* in which the aggregation number may vary from ~ 100 to ~ 1000 molecules. The aggregation occurs suddenly at a *critical micellar concentration* (*cmc*). In the micelles, the inter-actions between the paraffinic chains

$$\left(\diagup \begin{array}{c} C \\ \end{array} \diagdown \begin{array}{c} \\ C \end{array} \diagup \begin{array}{c} C \\ \end{array} \diagdown \begin{array}{c} \\ C \end{array} \diagup \begin{array}{c} C \\ \end{array} \diagdown \begin{array}{c} \\ C \end{array} \diagup \begin{array}{c} C \\ \end{array} \diagdown \begin{array}{c} \\ C \end{array} \diagup \begin{array}{c} C \\ \end{array} \diagdown \begin{array}{c} \\ C \end{array} \diagup \right)$$

and water are minimized by aggregation, whereas the interactions between the paraffinic chains by themselves are of the van der Waals type. The hydrophilic (charged) groups of the long-chain molecules are at the surface of the micellar-aggregate with water. Because of this location, the hydrophilic, or polar, heads of the chains interact strongly with water, and a *Helmholtz layer* of ionic groups, or charges of opposite sign, is present. In addition, a "*diffuse*" layer of mobile charge is formed opposite to the polar heads by the localization of the free counter-ions. This is further away as a diffuse (double) layer, formed with these counter-ions. The ions adsorbed from solution on the surface form the *Stern layer* (Stern, 1924). (See also Chap. 1 for additional details.)

One view is that micelle formation can be regarded as a *quasi*-phase-sepa-ration process, with the cmc being somewhat like a saturation limit of solubility. Cmc values are typically $\sim 10^{-2}$ to 10^{-3} mol/l, depending on the chain length. The sharp micellization at cmc occurs because of the multiparticle equilibrium condition:

$$n \cdot R{-}COO^{-} \overset{k_n}{\rightleftharpoons} (R{-}COO^{-})_n$$

$$k_n = a(R{-}COO^{-})_n / [aR{-}COO^{-}]^n$$

where the a's are *activities* (concentrations) of the monomers (long-chain non-polar molecules) of the micellar species. If this equation is solved for the activ-ity (concentration) of the micellar species $(R{-}COO^{-})_n$, the value of $[R{-}COO^{-}]$ (or activity) for $(R{-}COO^{-})_n$ is very sensitive to the monomer concentration (or activity) $[a(R{-}COO^{-})]$, since the latter is raised to the nth power (the aggregation number), which can be more than 100.

The morphology of micelles is also of some interest. Simple colloidal elec-trolytes form spherical micelles (Fig. 6-3), but depending on the molecule form-ing the micelle, rodlike or platelike (lamellar) micelles are also formed. The

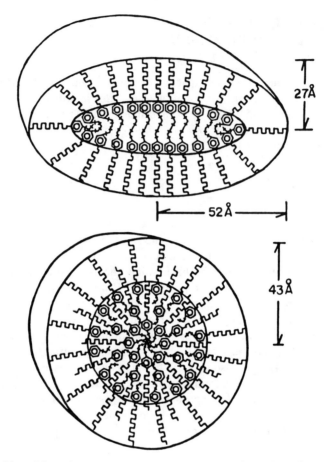

FIGURE 6-3.   Schematic representations of (*top*) the classical *oblate ellipsoid* model, and (*bottom*) the (nonclassical) spherical model for Triton X-100 micelles (from Shick, 1987, with permission).

most interesting and biologically significant example is the *lipid bilayer*, present in biomembranes (see, for example, Chap. 17 in Vol. II of this book).

CALCULATION OF FREE ENERGY AND ENTHALPY CHANGES

Reversible Processes in Foods and Biochemical Systems

The calculation of Gibbs free-energy changes starts with the selection of the states of the reversible process under consideration, either in a *closed* or an *open* thermodynamic system (see also Chap. 2 for the definitions of the relevant concepts). Thus, the *Gibbs free-energy change* is calculated as ΔG = Gibbs free energy in the final state *minus* Gibbs free energy in the initial state, or $\Delta G = G_{final} - G_{initial}$.

Under conditions of constant temperature and pressure, the *enthalpy change* is similarly calculated as $\Delta H = H_{final} - H_{initial}$.

As an example, consider the reaction of ATP hydrolysis

$$ATP \rightleftharpoons ADP + P_i$$

At pH = 7.0, the free-energy change, in the absence of Mg^{2+} ions, is calculated to be around -10 kcal/mole, whereas the "standard" free-energy change $\Delta G^{\dagger}$ is only about -7.0 kcal/mole for the reaction with all the reactants in their "standard" states. (The standard states are often labeled as $G°$, but this would imply a somewhat absolute means of identifying the standard states, whereas in practice the choice is one dictated by convenience only.) The actual ATP hydrolysis reaction is considerably more complex and involves several steps that are not discussed here. Catabolic processes can generate ATP that is then hydrolized by cellular processes requiring energy. In general, the reaction

$$nA + mB \rightleftharpoons pC + qD$$

proceeds with a *standard* free-energy change $\Delta G^{\dagger}_{reaction} = \Sigma\, G^{\dagger}_{products} - \Sigma\, G^{\dagger}_{reactants}$. (The standard free energy of a substance is defined for 1 mol under standard conditions of temperature and pressure that are constant.) In solution, the free energy of a substance A is related to its standard free energy as follows: $G_A = G^{\dagger}_A + RT \cdot \ln [A]$.

ELEMENTS OF THE DEBYE THEORY OF DILUTE SALT SOLUTIONS

Electrostatic Interactions

The interaction energy, E, between two charges q_1 and q_2, separated by a distance R_{12} in a medium of dielectric constant ϵ_0 is readily obtained from *Coulomb's law* by multiplying the electrostatic force by the distance R_{12}. The result is

$$E = \frac{332}{\epsilon_0} \cdot \frac{q_1 \cdot q_2}{R_{12}} \tag{6-2}$$

where the charges are expressed in units of electron charge, e_0, E is expressed in kcal/mol, and R_{12} is in Å.

In the case of partial charges, one often has such a charge as part of an electric dipole moment, μ. In general, for two (point) electric *dipole* moments μ_1 and μ_2 separated by a distance $\vec{R}_{12}$, the *interaction energy* is

$$E = \frac{332}{\epsilon} \cdot \left[\frac{\mu_1 \cdot \mu_2}{R_{12}^3} - 3(\mu_1 \cdot \vec{R}_{12}) \cdot \frac{\mu_2 \cdot \vec{R}_{12}}{R_{12}^5} \right] \qquad (6\text{-}3)$$

The unit of dipole moment is the *Debye*:

$$1 \text{ debye} = 1 \text{ D} = 0.208 e_0 \cdot \overset{\circ}{A}$$

The interaction energy for two dipole moments of 1 D that are colinear, pointing toward each other, and separated by 5 Å in a medium of dielectric constant $\epsilon = 4$, is 1.32 kcal/mol. If the two dipoles are antiparallel, but not colinear, the interaction energy would be -0.66 kcal/mol (attractive). The dielectric constant value $\epsilon = 4$ is the macroscopic value for amide polymers, whereas liquid water at 20°C has $\epsilon = 81$. Although neutral molecules have zero net charge, such molecules may still have nonzero dipole, or higher multipole, moments. These dipoles and multipoles contribute to the total interaction energy by a value determined from equation (6-3). For a large biopolymer, such as a protein, the calculation of all the electrostatic interactions would be very time-consuming because of the large number of atoms involved, as well as the long range of the electrostatic interactions. In a first approximation, however, by ignoring charge fluctuations in a polyelectrolyte system, one can consider only dipoles and multipoles in order to calculate the interaction energy with equation (6-2). In such cases the energy calculation can be restricted to the interactions between neighbors at short distances. For example, in β-pleated sheets in proteins the adjacent dipoles are antiparallel, and their electrical fields cancel each other at long distances. On the other hand, in α-helices of polypeptide chains the dipoles form lines that give net charges only at both ends; even this electrostatic contribution is small when it is compared with the total binding energy for such α-helices. The contributions of *charge fluctuations* and *fluctuating dipoles* are discussed later in this chapter, using the *Kirkwood and Shumaker theory*.

Debye–Hückel Theory (or Model) of Dilute Electrolyte Solutions

The simplest thermodynamic theory of ideal gases was briefly discussed in Chapter 2, and earlier in this chapter the van der Waals interactions that significantly modify such ideal behavior were described (e.g., eq. (6-1) and Table

6-5). The concept of electrical double layers surrounding particles in aqueous dispersions containing electrolytes was also introduced in Chapter 1. The early double-layer models could be considered as analogous to a parallel condenser on a molecular scale. Gouy (1910) and Chapman (1913) indicated the short-comings of this oversimplified model, paving the way for the *Debye–Hückel theory* (or model), which is concerned with the electrical potential calculation for the water molecules surrounding salt ions in dilute solutions. In general, systems consisting of discrete charges (ions) placed in dielectric media are subject to Poisson's equation:

$$\nabla(\epsilon(r) \cdot \nabla\psi(r)) = -4\pi\rho(r) \tag{6-4}$$

which takes the simpler form

$$\nabla^2\psi = \Delta\psi = -\frac{4\pi\rho}{\epsilon} \tag{6-5}$$

for a *homogeneous medium* or solvent of dielectric constant ϵ, in which the *charge density* ρ is related to the ion concentration as follows:

$$\rho = \sum Z_i \cdot e \cdot n_i$$

The Coulomb interaction between the ions and the solid surface of the dispersed particle (or colloid) is perturbed, or modulated, by thermal motions. Therefore, the local concentration of the ions, $n_\pm$, will be statistically distributed according to *Boltzmann's law*

$$n_\pm = n \cdot \exp\left(\mp z_\pm \cdot \frac{e\psi}{kT}\right) \tag{6-6}$$

where n is the bulk, average concentration of the ions at long distances from the particle surface, T is the absolute temperature, and k is the Boltzmann constant. By combining equations (6-4) and (6-5) one obtains the *Poisson-Boltzmann equation* for an electrolyte solution:

$$\frac{d^2\psi}{dx^2} = -\frac{4\pi e}{\epsilon} \cdot \left[nZ_+ \cdot \exp\left(-\frac{Z_- e\Psi}{kT}\right) - nZ_- \cdot \exp\left(\frac{Z_- e\Psi}{kT}\right)\right] \tag{6-7}$$

written for a direction x, arbitrarily chosen in the liquid. Because of the spherical symmetry of the random ion distribution in solution, it is advantageous to employ polar (spherical) coordinates and rewrite equation (6-7) in the equivalent but simpler form

$$\frac{d^2\Psi}{dx^2} = \frac{1}{r^2} \cdot \frac{\partial}{\partial r}\left(r^2 \cdot \frac{\partial\Psi}{\partial r}\right)$$

$$= -\frac{4\pi e}{\epsilon} \cdot \left(nZ_+ \exp\frac{-Z_+ e\Psi}{kT} - nZ_- \exp\frac{Z_- e\Psi}{kT}\right) \qquad (6\text{-}8)$$

where the electrical field is $|E| = -d\psi/dr$. If the radius of the dispersed (or colloidal) particle is a, then one has the *boundary condition* $\psi = \psi_a(r = a)$. Since the Gouy–Chapman model of the electrical double layer assumes that the ion statistical distribution begins precisely at the particle surface, one also has that $\psi = \psi_0$ (Fig. 6-4a). Such an assumption yields too *large* an ion concentra-

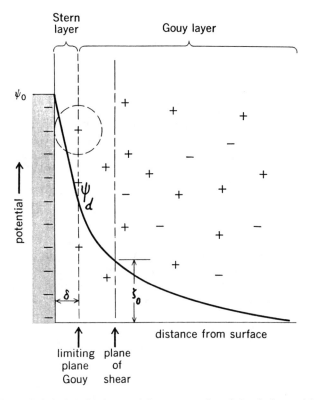

FIGURE 6-4a.   Electrical double layer and the representation of electrical potential according to the Debye–Hückel and Stern-layer theories. The ψ_0 is the potential at the particle surface, and ψ_d is the potential at the surface of the Stern layer (from Parker, 1983, with permission).

tion at the particle surface, because the model assumes *point* charges (of zero volume). This was corrected by the Stern (1924) and Grahame (1947) models, which consider ions of *finite* radii that are adsorbed onto the large particle surface forming the *Stern layer* of adsorbed ions (Fig. 6-4b) of finite thickness, d; beyond the Stern layer the diffuse ion region begins (or Gouy layer) to which the Gouy–Chapman parallel-plate condenser model would apply. For small potentials, that is, for $Z\psi \ll 25$ mV (which is likely to occur far away into the diffuse layer, beyond the plane of shear, and also only in dilute solutions), $Z_i e\psi < kT$ and equation (6-8) has a relatively simple (Debye–Hückel) solution for the potential function, because the exponentials in equation (6-8) can be expanded as *power series* $e^{-W/kT} => 1 - W/kT$. If only the first-order terms in

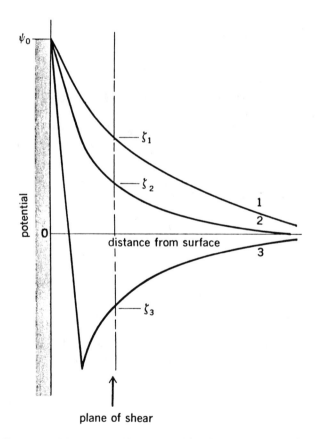

FIGURE 6-4b.   Potentials across the Stern layer and the plane of shear in the presence of adsorbed ions onto the particle surface. Note the reversal of the zeta potential ζ_0 caused by the adsorption of the ions (from Parker, 1983, with permission).

$(z_i e\psi/kT)$ are retained, the charge density ρ can be written as

$$\rho = \sum_i (Z_i \cdot n_{i0} \cdot e^2 \Psi)/kT = \epsilon \kappa^2 \Psi \tag{6-9}$$

where

$$\kappa^2 = e^2 \left(\sum_i Z_i^2 \cdot n_{i0} \right) \Big/ \epsilon kT \tag{6-10}$$

Replacing ρ in equation (6-8) with the expression from equation (6-9) yields

$$\frac{d^2\psi}{dx^2} = \kappa^2 \psi \tag{6-11}$$

which has the simple Debye–Hückel solution

$$\Psi = \Psi_0 \cdot \exp(-\kappa x) \tag{6-12}$$

with the boundary conditions $\psi \rightarrow \psi_0$ for $x \rightarrow 0$ and $\psi \rightarrow 0$ for $x \rightarrow \infty$. The exponential decay of ψ with distance from the surface is illustrated in Figure 6-4. Note that κ^{-1} has the units of length, and is often called the "thickness" of the double layers. The parameter κ is concentration dependent and can also be expressed as

$$\kappa = \left(\frac{1000 l^2 N_A}{\epsilon kT} \cdot \sum_i Z_i^2 \cdot M_i \right)^{1/2} \tag{6-13}$$

where the summation within the parentheses is twice the *ionic strength* of the solution.

SOLUTE ACTIVITY AND CHEMICAL POTENTIAL

Hückel suggested an empirical treatment of the nonideal behavior of electrolyte solutions and ion hydration effects based on the Debye–Hückel model introduced earlier, in terms of the ion activity,

$$a_i = \gamma c_i \tag{6-14}$$

where a_i is the *ion activity*, γ is the *activity coefficient*, and c_i is the concentration. Hückel showed that the *mean activity coefficient* $\gamma_\pm$ can be expressed in terms of the parameter κ defined in equations (6-10) and (6-13):

$$\ln \gamma_{\pm} = \left(\frac{-Z^2 e^2}{2\epsilon kT}\right)\left(\frac{\kappa \cdot \mathring{a}}{1 + \kappa \cdot \mathring{a}}\right) + f(\kappa) \tag{6-15}$$

where the Debye *reciprocal length*, κ, is considered to be the same as the reciprocal radius of the *ionic atmosphere* and $f(\kappa)$ has the value $C \cdot (c_i z_i^2)$, where C is a positive constant that depends on the *dielectric decrement*, δ, caused by the ions in water (related to the orientation of water molecules in the ion hydration shell(s)). This effect of the dielectric decrement of salt solutions on the salt mean activity coefficient is illustrated in Figure 6-5. Note that $\ln \gamma_{\pm}$ takes larger values than those predicted by the Debye–Hückel model (eq. (6-12)) when C is positive, as was also found experimentally (Fig. 6-6a and b). At sufficiently high salt concentrations $\gamma_{\pm} > 1.0$ ($\ln \gamma_{\pm} > 0$), as in the upper part of Figure 6-5. In Equation (6-15), $\mathring{a}$ is the *distance of closest approach* between ions.

The *chemical potential* of a salt in solution is defined as

$$\mu_2 = \left(\frac{\partial \epsilon}{\partial n_2}\right)_{T,P,n_1,\cdots} \tag{6-16}$$

where ϵ is the free energy of the system and n_2 is the number of moles of salt. Under ideal conditions, the *ideal chemical potential* of the solute (salt) is

$$\mu_{\text{ideal}} = \mu_2^{\dagger} + RT \ln c_2 \tag{6-17}$$

where $\mu^{\dagger}$ (or μ^*) is the "*standard*" chemical potential, R is the universal gas constant, and c_2 is the salt concentration. For a *nonideal* solution, there is a *nonideal free-energy* contribution $RT \ln \gamma_2$ (or $RT \ln \gamma_{\pm}$) to the chemical potential of the salt in solution. The mean activity coefficient, $\gamma_{\pm}$, for cations and anions of the salt is defined as

$$\gamma_{\pm} = \frac{\gamma_+ \cdot \nu_+ + \nu_- \cdot \gamma_-}{\nu_+ + \nu_-} \tag{6-18}$$

where ν_+ and ν_- are, respectively, the numbers of cations and anions in solution. Therefore, under nonideal conditions

$$\mu_{\text{nonideal}} = \mu^{\dagger} + RT \ln c_2 + RT \ln \gamma_{\pm}$$

$$\mu_{\text{nonideal}} = \mu_2^{\dagger} + RT \ln (\gamma_{\mp} c_2) = \mu^{\dagger} + RT \ln a_2 \tag{6-19}$$

where a_2 is the solute (salt) activity.

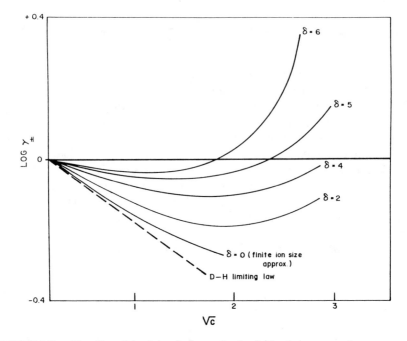

FIGURE 6-5. The effect of the dielectric decrement, δ, of salt solutions, treated as a parameter, on the activity coefficients of the ions (from Conway, 1981, with permission).

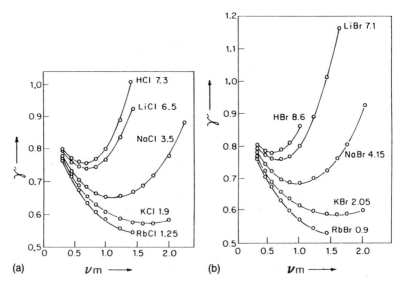

FIGURE 6-6. Variation of $\gamma_\pm$ with concentration for electrolyte solutions at high ionic strengths for (a) alkali chlorides, HCl and (b) alkali earth bromides (Stokes and Robinson, 1948; from Conway, 1981, with permission).

In dilute solutions, the $RT \ln \gamma_2$ or $RT \ln \gamma_\pm$ free-energy contribution to the chemical potential of the salt in solution, μ, is caused primarily by the *long-range Coulomb* forces between ions. The Debye–Hückel model, based on the *low-potential, spherical approximation* predicts a negative value for $RT \ln \gamma$, that is, $\gamma < 1.0$. The hydration effects are then taken into account with *Hückel's approximation* (eq. (6-15)), which introduces the distance-of-closest approach, å, between ions in solution, which is treated as a *parameter*. Usually, å represents an empirical *"distance of interionic contact between the hydrated ions"* (Conway, 1981).

Note that aqueous solutions of amino acids cannot be treated by applying the Debye–Hückel directly to individual amino acid charges (Slater and Kirkwood, 1931) because they form *zwitterions* (see also formula (F-4-19)) that have appreciable *dipole moments* in aqueous solutions. The properties of water (Kirkwood, 1932) and dilute amino acid solutions with and without added ions were first treated by Kirkwood (1934), who applied the Debye–Hückel theory to the amino acid dipoles, instead of single charges. Kirkwood also suggested the use of Keesom's theory of dipole gases (Keesom, 1921) at the higher amino acid/salt concentrations, in order to account for the *mutual electrostatic interaction* of the amino acid zwitterions and/or salt ions. Subsequently, Tanford (1957, 1961) showed that the calculation of *charge–charge interactions* of amino acid groups in proteins with the Kirkwood procedure (1934) requires the placement of such charges at a *constant* depth, d, of about 1 Å from the protein surface, toward the protein interior.

Gibbs–Duhem Equation

The *chemical potential of the solvent* in a salt solution, μ_1, is defined as

$$\mu_1 = \left(\frac{\partial \epsilon}{\partial n_1} \right)_{T,P,n_2,\,\cdots} \tag{6-20}$$

and can be also expressed under ideal conditions as

$$\mu_1 = \mu_1^\dagger + RT \ln c_1$$

where n_1 is the number of moles of solvent, c_1 is the solvent concentration, and $\mu_1^\dagger$ (or μ_1°) is the "standard" chemical potential of the solvent. In *nonideal* situations, as in the case for medium to high salt concentration,

$$\mu_1 = \mu_1^\dagger + RT \ln a_1 \tag{6-21}$$

where $a_1 = \gamma_1 \cdot c_1$ is the *solvent activity* and γ_1 is the *solvent activity* coefficient. For water as a solvent, one has

$$\mu_w = \mu_w^\dagger + RT \ln a_w \qquad (6\text{-}22)$$

and

$$a_w = \gamma_w \cdot c_w \qquad (6\text{-}23)$$

where a_w is the *water activity* and γ_w is the activity coefficient of water.

Consider the important fact that the chemical potentials of the solute (eq. (6-16)) and of the solvent (eq. (6-20)) are *not* independent of each other; that is, when the solute chemical potential increases, that of the solvent must decrease proportionally. This is expressed in a precise form by the *Gibbs–Duhem equation:*

$$n_1 \, d\mu_1 + n_2 \, d\mu_2 = 0.0$$

for a solution of n_2 molecules of solute "2" (salt) in n_1 moles of solvent "1," (such as water). This can be also expressed as

$$n_1 \, RTd \,(\ln a_1) + n_2 RTd \,(\ln n_2) = 0 \qquad n_2 d \,(\ln a_2) = -n_1 d \,(\ln a_1)$$

$$(6\text{-}24)$$

This relationship indicates that at high salt concentrations in water, for example, water activity will decrease substantially because a significant fraction of water will be bound by the cations and/or anions. Such ion-bound water has diminished activity and is less available as a solvent, which means there is a higher apparent ion concentration, or salt activity. As the solvent activity decreases, the solute mean activity coefficient $\gamma_\pm$ increases. The effect of hydration on γ is specific to the type of ion and is described in the Hückel approach through the parameter $\mathring{a}$ of closest distance (eq. (6-16)). Stokes and Robinson (1948, 1949) introduced an improved theory of hydration effects on the ionic activity coefficients by considering well-defined hydration numbers for the ions in solution. Their result is

$$\ln \gamma_\pm = -A|Z_+ Z_-| \sqrt{I}/(1 + \mathring{a}B\sqrt{I}) - (n/\nu) \ln a_1$$

$$- \ln [1.0 + 0.001m \cdot M_1(\nu - n)] \qquad (6\text{-}25)$$

where a_1 is the solvent activity and can be obtained from vapor pressure measurements; I is the ionic strength, $(1/2) \, \Sigma_i \, C_i Z_i^2$; n is the hydration number for the salt in solution; A is a constant; $\overset{\circ}{a}$ is the contact radius of the solvated ions; and $\nu = \nu^+ + \nu^-$ is the number of moles of cations plus that of anions from the same salt.

NONIDEAL BEHAVIOR OF DISPERSIONS

In this section we discuss a few examples of nonideal behavior in dispersions. Virial coefficients, charge fluctuations, the *Kirkwood–Schumaker theory*, osmotic pressure, and light-scattering observations of protein–protein interactions are also briefly discussed.

Charge Fluctuations and the Role of London Dispersion Forces in Aggregation of Particles

Most hydrophobic dispersions in water are *unstable*, unless an emulsifier is added (see Chap. 1 in this volume). Upon addition of an electrolyte to such dispersions, aggregation (or *flocculation*) occurs, caused by the reduction, or screening, of the repulsive interactions between the dispersed particles and the effects of hydrophobic interactions among the particles. The attractive forces able to provide the necessary energy to such an aggregation process are the *London dispersion forces*, which result from the charge fluctuation in an atom caused by the electron motion within the atom orbitals. Such charge fluctuations produce transient dipoles of opposite orientation in neighbor atoms that would flip-flop together, thus always causing attraction between the neighbor atoms. For particle separations or diameters larger than ~ 100 Å, these attractive forces have decreased substantially.

The reduction of the potential energy of repulsion between two spherical particles of radius a for various values of κa, caused by the addition of a symmetric 1:1 electrolyte, is illustrated in Figure 6-7 (according to Kruyt, 1952).

The London dispersion forces of attraction can be readily estimated in vacuum for either two large parallel plates or two large spherical particles. For two infinitely long colloidal plates of thickness δ placed at a distance from each other in vacuum, the attractive potential is

$$V_A = -\frac{A}{48\pi}\left[\frac{1}{0.5r^2} + \frac{1}{(0.5r + \delta)^2} - \frac{2}{(0.5r + 0.5\delta)^2}\right] \qquad (6\text{-}26)$$

when $r \gg \delta$. This expression can be approximated by

$$V_A^* = -\frac{A}{12\pi r^2} \qquad (6\text{-}27)$$

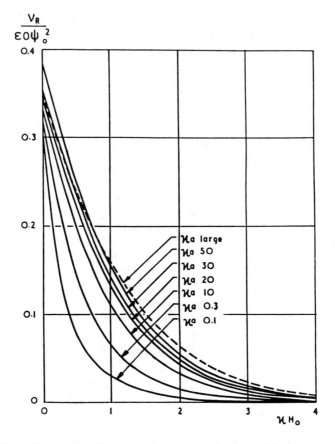

FIGURE 6-7. The variation of the potential-energy curve for the repulsion between two spherical particles, with the value of (κa) treated as a parameter (from Kruyt, 1952, with permission).

where A is the *Hamaker constant* determined as

$$A = (3/4)\pi^2 q^2 \cdot h\nu \cdot \alpha^2 \qquad (6\text{-}28)$$

with q being the number of atoms in 1 cm^3 of the particle material. In water the Hamaker constant is expressed as

$$A = A_{00} + A_{11} - 2A_{10} \qquad (6\text{-}29)$$

where A_{00} is the force between water molecules, A_{10} is the force between the plates and water, and A_{11} is the force between the two plates. The value A_{00} for liquid water was reported to be 0.6×10^{-12} erg by Slater and Kirkwood (1931).

For two spheres of equal radius A and a distance r between their center, Hamaker (1937) calculated the attractive force as

$$V_A = -\frac{A}{6} \cdot \left[\frac{2a^2}{r^2 - 4a^2} + \frac{2a^2}{r^2} + \ln \frac{r^2 - 4a^2}{r^2} \right] \qquad (6\text{-}30)$$

Virial Coefficients of Water and Protein Solutions in Water

A case of special interest is that of protein molecules in aqueous solutions. Although many groups in a protein are hydrophobic, protein solutions are often stable up to high electrolyte concentrations if the protein is not *predominantly hydrophobic* (Schachman, 1959; Scatchard, 1943). The forces responsible for the stability of such protein dispersions also cause characteristic *nonideal* behavior in such systems. To begin with, let us consider the nonideal behavior of low-density water vapor. The thermodynamic equation of the state of water for pairwise-additive potentials is

$$\frac{pV}{NkT} = 1 - \frac{1}{8NkT} \int dx_1 \cdot \int dx_2 [r_{12} \cdot \nabla_{12} v(x_1, x_2)] \rho^{(2)}(x_1, x_2) \qquad (6\text{-}31)$$

where $r_{12} = r_2 - r_1$, $\rho^{(2)}(x_1, x_2)$ is the *pair-distribution function* for water, $v(x_i, x_j)$ is the pair potential for water molecules i and j, and $\rho(x_1, x_2) = (N/v)^2 g^{(2)}(x_1, x_2)$, where $g^{(2)}$ is a correlation function for the positions of the pair of water molecules (according to Ben-Naim and Stillinger, 1972). Because of the low density of the water-vapor system, the preceding equation of state (6-31) can be expanded into a *density*, power series ($\rho \to 0$),

$$\frac{pV}{NkT} = 1.0 + B(T) \cdot \left(\frac{N}{V}\right) + C(T) \cdot \left(\frac{N}{V}\right)^2$$

$$+ D(T) \cdot \left(\frac{N}{V}\right)^3 + \cdots \text{ high-order terms} \qquad (6\text{-}32)$$

The coefficients of this series expansion, $B(T)$, $C(T)$, $D(T)$, and so on, are called, respectively, the second, third, fourth, $\cdots$, *virial coefficients*, and represent interactions among *clusters* of increasing numbers of water molecules (couples, or *pairs*, in the case of $B(T)$, *triplets* in the case of $C(T)$, etc.). Thus, they also represent deviations from the *ideal random* gas configuration, since these correspond to a decrease in the entropy of the system in comparison with a random configuration.

The *orientational correlation function* of water, g_K, is related to the pair-correlation function $g^{(2)}(\mathbf{x}_1, \mathbf{x}_2)$ through the equation

$$g_K = 1.0 + (N/8\pi^2 \cdot V) \int dx_2 \, (\mathbf{b}_1 \cdot \mathbf{b}_2) g^{(2)}(\mathbf{x}_1, \mathbf{x}_2) \qquad (6\text{-}33)$$

The function g_K gives the average cosine of the angle between permanent dipole moment directions b_1 and b_2 for two neighbor water molecules. The orientational correlation function g_K appears also in the expression of the static dielectric constant, ϵ_0, of water calculated with *Kirkwood's theory of polar dielectrics* (Kirkwood, 1939):

$$\frac{(\epsilon_0 - 1)(2\epsilon_0 + 1)}{3\epsilon_0} = \frac{4\pi N}{V} \cdot \left(\frac{\alpha + \mu_d^2 \cdot g_K}{3kT}\right) \qquad (6\text{-}34)$$

where α is the *molecular polarizability of water* and μ_0 is the *permanent dipole moment* of water.

The chemical potential μ_w of water can be calculated by assuming that the *partially coupled* water molecule "1" interacts with its neighbors with a potential $v(x_i, x_j; \xi)$, which is *fully coupled* for a neighbor with a value $\xi = 1$, so that $v(x_i, x_j; \xi = 1.0) = v(\mathbf{x}_i, \mathbf{x}_j)$, and is *completely decoupled* from the other neighbors that have $\xi = 0$, such that $v(\mathbf{x}_i, \mathbf{x}_j; \xi = 0.0) \equiv 0.0$ for all the fully decoupled neighbors. The amount of *reversible* mechanical work required to switch on $v(i, j; \xi)$ (that is, the amount required to increase the potential from 0.0 to $v(\mathbf{x}_i, \mathbf{x}_j)$ by increasing ξ from 0.0 to 1.0) gives the water chemical potential

$$\mu = \mu^{(\dagger)} + kT \ln \frac{N}{V} + \frac{1}{N} \cdot \int_0^1 d\xi \int dx_1 \frac{\partial v(\mathbf{x}_1, \mathbf{x}_2; \xi)}{\partial \xi} \rho^{(2)}(\mathbf{x}_1, \mathbf{x}_2; \xi) \qquad (6\text{-}35)$$

where $\mu^\dagger$ is the *standard chemical potential of water*.

Note that the first two terms in this expression represent the *ideal gas* contributions, while the third, integral term accounts for the interactions that cause the *nonideal* behavior of water. This expression is supposed to hold even at high water-vapor densities and in the liquid (Ben-Naim and Stillinger, 1972); for the low-density water vapor, equation (6-35) may also be expanded in series, resulting in an expression similar to equation (6-32) for water activity in terms of the *virial expansion coefficients*, $B(T)$, $C(T)$, $D(T)$, and so on.

We now return to a consideration of the nonideal behavior of protein solutions in water. Forces between protein molecules in stable solutions, or dispersions, are mainly *electrostatic* in origin, as seen from the fact that the ionic strength has a marked effect on the solubility of proteins in aqueous solutions (Scatchard, 1943; Scatchard and Kirkwood, 1932). Some specific examples were given in Chapter 1. Other thermodynamic parameters are also strongly affected by the electrostatic forces between protein molecules in solution. At pH values different from the isoionic points of specific proteins, such effects are primarily caused by a simple *nonspecific* Coulomb force, determined by the *net* electric charge of such proteins. On the other hand, when the proteins are at their isoionic point, there are other, structure-sensitive electrostatic forces that come into play. Such forces can be attributed to both *static* configurations of electrical charge and *electrical changes* that result in dipole multipole moments and *fluctuations* in the charge distribution (as well as charge) associated with fluctuations in the number and configuration of protons bound to the protein side chains (Kirkwood and Shumaker, 1952a, 1952b). Among such groups are those of His, Glu, Asp, Arg, and Lys. Linderstrøm-Lang (1924) have shown that fluctuations in the net charge of the protein contribute to the interaction between proteins and small ions of dissolved electrolytes in the same solution. Kirkwood and Shumaker (1952b) postulated that, at the isoionic point, there is a long-range *attractive* force between protein molecules with a potential that decreases asymptotically with the distance r between the protein molecules, as $(1/r^2)$, in the absence of screening by electrolytes (no salt added). In the presence of charge screening by added electrolytes, the range of this interaction is reduced by an exponential factor, $e^{-\kappa \cdot r}$, where κ is the Debye–Hückel parameter (Kirkwood and Shumaker, 1952b). The expression derived by Kirkwood and Shumaker for this interaction potential, $W_{11}(r)$, is

$$W_{11}(r) = [\langle \Delta q_{(1)}^2 \rangle_{AV} \cdot \langle \Delta q_{(2)}^2 \rangle_{AV} / 2D^2 r^2 \cdot kT] \cdot e^{-2\kappa(R-a_{12})}/(1 + \kappa a_{12})^2$$

(6-36)

where $\kappa^2 = \kappa_0^2 + \kappa_1^2$, with $\kappa_0^2 = (4\pi Ne^2/100DkT) \Sigma_i c_i z_j^2$

$$\kappa_1^2 = (4\pi N/100DkT) [(\langle \Delta q_{(1)}^2 \rangle_{AV} \cdot C_1/M_1) + (\langle \Delta q_{(2)}^2 \rangle_{AV} \cdot C_2/M_2)]$$

(6-37)

In the preceding equations, a_{12} is the sum of the radii of the two protein molecules of species 1 and 2, κ_0^2 is proportional to the ionic strength of the added electrolytes, the protein concentrations C_1 and C_2 are expressed in grams per 100 ml in solution, and the molecular weights of the protein species 1 and 2 are M_1 and M_2, respectively. The total charge fluctuations of the two protein molecules are $\langle \Delta q_{(1)}^2 \rangle_{AV}$ and $\langle \Delta q_{(2)}^2 \rangle_{AV}$, respectively.

For a single protein species (1) at its isoionic point, the effect of the protein–protein interactions on the chemical potential is given by the "excess" chemical potential, μ_1^e in the following expression

$$\mu_1 = kT \ln C_1 + \mu_1^e + \mu_1^\dagger(T, p) \tag{6-38}$$

Note that

$$\left(\frac{\partial \mu_1^e}{\partial c_1}\right)_{T,p,\kappa 0} \cdot \left(\frac{100 M_1}{NkT}\right) = G_{10} - G_{11} \tag{6-39}$$

where the aqueous solution has been assigned label 0 and the protein the subscript label 1.

In the preceding expression,

$$G_{10} = 4\pi \int_0^\infty R^2 [g_{10}(r) - 1.0] \, dr \tag{6-40}$$

and

$$G_{11} = 4\pi \int_0^\infty R^2 [g_{11}(r) - 1.0] \, dr \tag{6-41}$$

where $g_{10}(r)$ is the radial (pair) distribution function of a pair consisting of a protein and a water-solvent molecule, whereas $g_{11}(r)$ is the pair-distribution function for the protein molecules by themselves. The potentials resulting from the average force, $W_{11}(r)$ and $W_{10}(r)$, are related to these pair-distribution functions as follows

$$g_{10}(r) = e^{-\beta W_{10}(r)} \quad \text{and} \quad g_{11}(r) = e^{-\beta W_{11}(r)} \tag{6-42}$$

The integrals in equations (6-40) and (6-41) include a *co-volume* term corresponding to the overlap of the *excluded volumes* of the molecular pairs (1–0) and (1–1), respectively. By assuming that liquid water plays only the role of a dielectric continuum *outside* the excluded volume of the protein, $W_{11}(r)$ also can be expanded in series, and one has that

$$\left(\frac{\partial \mu_1}{\partial c_1}\right)_{T,p,\kappa 0} = 2B_0 - \pi N \langle \Delta q_{(1)}^2 \rangle_{AV} / [(DkT)^2 \cdot \kappa(1 + \kappa a)^2] \tag{6-43}$$

The co-volume contribution is $B_0 = 7\pi N a^3 / 12$, and a is the protein molecule diameter (for a globular protein considered as a sphere). The *total charge fluctuation* $\langle \Delta q_{(1)}^2 \rangle_{AV}$ of the protein molecule is given by the expression

$$\langle \Delta q_{(1)}^2 \rangle_{AV} = e^2 \sum_{i=1}^{\nu_1} 1 / [2 + K_1^{(1)} / [H^+] + [H^+] / K_1^{(2)}] \tag{6-44}$$

Assuming that the ionic strength is primarily determined by the added electrolyte, equation (6-43) gives by integration:

$$\mu_1^e/(kT) = 2BC_1 \tag{6-45}$$

with

$$B = (1/100M_1)\{B_0 - \pi N \langle \Delta q_{(1)}^2 \rangle_{AV}/[2(DkT)^2 \kappa_0 (1 + \kappa_0 a)^2]\} \tag{6-46}$$

(Kirkwood and Shumaker, 1952b).

The *virial coefficient B* is the same as the corresponding virial coefficient in the expansion of the osmotic pressure in powers of C_1, with an expression similar to equation (6-32).

At high ionic strengths (that is, in the presence of charge screening of the protein molecules), the co-volume contribution to the interaction coefficient B in equation (6-46) dominates, resulting in a *positive* value for the excess chemical potential, μ_1^e. On the other hand, at very low ionic strengths, the second term in equation (6-43), which represents the charge fluctuation caused by the proton fluctuations among various protein sites, becomes dominant. The latter term results in a negative value of the excess chemical potential, μ_1^e (Kirkwood and Shumaker, 1952a, 1952b). For example, the coefficient B of serum albumin in water at 300 K and at the isoionic point is $B = (4.3 - 1.2/\sqrt{I})/1000$, where I is the ionic strength of the added, dissolved salt. For an ionic strength of 0.001 the fluctuation term is approximately ten times as large as the co-volume, B_0, term. The values of the excess chemical potential for 2% to 5% serum albumin solutions are therefore *negative*, which corresponds to a substantial *attractive* force between the proteins. This *long-range attractive force* arises from *fluctuations in the protein charge*, as well as the fluctuation in the *configuration* of mobile protons. Note that equation (6-36) differs from that of Hamaker for the attraction between spheres by London dispersion forces (eq. (6-30)), although both have terms in $(1/r^2)$. Note also that the meaning and values of the coefficients, or constants, multiplying the $(1/r^2)$ terms are different.

Timasheff et al. (1957), Timasheff and Kronman (1959), and Timasheff and Coleman (1960) were able to provide experimental evidence for the *charge fluctuation* mechanism of Kirkwood and Shumaker (1952b), which was discussed earlier, by measuring the light scattering in very dilute isoionic serum albumin solutions. In the case of very dilute isoionic protein solutions, the excess chemical potential in equation (6-45) can be simplified through *binomial expansion* of the term $1/(1.0 + \kappa a)^2$. The result of such an expansion is a power series in $C_1^{1/2}$ for the light-scattering intensity function $(H_1 C/\Delta\tau)$, whose coefficients were those from equation (6-46) of Kirkwood and Shumaker (1952b). The light-scattering data for isoionic serum albumin solutions gave a *linear* plot against $C_1^{1/2}$, which is the square root of the protein concentration. The fitted curve also included a linear term in C_1 whose coefficient was a *charge-fluctuation parameter* derived from the charge-fluctuation coefficient B (eq. (6-47)), and the excluded volume, B_0, as well as the dipole and multipole coefficients ($2B'$), were also included in the linear term in protein concentration, C_1. Addition of NaCl to the isoionic BSA solutions resulted in linear plots of the light-scattering data against protein concentration for NaCl concentrations higher than 1 mM. The slope of such plots increased with the NaCl concentration up to 150 mM as a result of the *electrostatic repulsion* between the protein surfaces caused by the presence of *bound chloride* ions. This electrostatic repulsion became stronger than the attractive force caused by charge fluctuations at NaCl concentrations of about 2 to 5 mM, leading to a *positive value* for the second virial coefficient (B). Subsequently, Vilker et al. (1981) measured the protein concentration dependence of the osmotic pressure for BSA in the con-

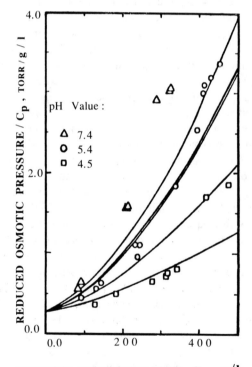

FIGURE 6-8.   Reduced osmotic pressure variation with the concentration of albumin solutions; C_P; the effects of molecular shape and size on the value of the calculated excluded volume contribution to the virial coefficient are illustrated (full lines from Vilker et al., 1981, with permission).

centrated range, and also studied the pH dependence in the presence of added salt (0.15 M NaCl). They were also able to fit their osmotic pressure data with the semiempirical equation

$$\pi = RT\{2[(ZC_1/2M_1)^2 + m_s^2]^{1/2} - 2m_s\} + (RT/M_1)(C_1 + A_2 C_1^2 + A_3 C_1^3)$$

$$(6\text{-}47)$$

The first term was chosen to represent the *Donnan effect*, whereas protein-protein, ion-ion, and protein-ion/water interactions were represented by the second term in equation (6-47), through the virial coefficients A_2 and A_3. The experimental data and the fitting reported by Vilker et al. (1981) are shown in Figure 6-8. In agreement with Timasheff et al. (1957), Vilker et al. (1981) also reported *positive values* of the second virial coefficient in the presence of 150-mM NaCl at pH values of 4.5, 5.4, and 7.4 (that is, *repulsion* between proteins). The *intermolecular potential* functions that are considered appear in Table 6-7, and the calculated pair-potential energy functions for the BSA so-

TABLE 6-7   Contribution to the Intermolecular Potential Function

Type	Designation	Unscreened Potential Function, $W*$	Screening Factor, ξ
Electrostatic			
Charge–charge	$W^{q,q}$	$+\dfrac{(Ze)^2}{\epsilon r}$	$\dfrac{g e^{-\kappa(r-2a)}}{(1+\kappa a)^2}$
Charge–dipole	$W^{q,\mu}$	$-\dfrac{2}{3}\dfrac{(Ze)^2\mu^2}{\epsilon^2 kTr^4}$	$\left\{\dfrac{3(1-\kappa r)e^{-\kappa(r-2a)}}{(1+\kappa a)[2+2\kappa a+(\kappa a)^2+(1+\kappa a)\epsilon_s/\epsilon]}\right\}^2$
Dipole–dipole	$W^{\mu,\mu}$	$-\dfrac{2}{3}\dfrac{\mu^4}{kT\epsilon^2 r^6}$	$\left\{\dfrac{3[2+2\kappa r+(\kappa r)^2]e^{-\kappa(r-a)}}{2+2\kappa a+(\kappa a)^2+(1+\kappa a)\epsilon_s/\epsilon}\right\}^2$
Charge fluctuation	$W^{\Delta q,\Delta q}$	$-\dfrac{\langle Z^2\rangle^2 e^4}{2\epsilon^2 kTr^2}$	$\dfrac{e^{-2\kappa(r-2a)}}{(1+2\kappa a)^2}$
Dispersion	$W^{\alpha,\alpha}$	$-\dfrac{A}{6}\left[\dfrac{2}{s^2}+\dfrac{2}{s^2-4}+\ln\left(\dfrac{s^2-4}{s^2}\right)\right]$	

where
$$A = (A_p^{1/2} - A_s^{1/2})^2$$

$$A_1 = \pi^2\left(\dfrac{\rho_1 N_A}{M_1}\right)^2 \dfrac{3}{4} h\nu_{0i}\alpha_i^2 \quad (i = p, s)$$

$$s = \dfrac{r}{a}$$

Source: Data from Vilker et al., 1981.

lutions are shown in Table 6-8 and Figure 6-9. Note that these *pair-potential* functions include only the *repulsive charge–charge* interactions (previously discussed by Timasheff et al., 1957) and the *attractive, London dispersion* interactions (Hamaker, 1937). This approach failed to fit the BSA data at pH 7.4, which was far from the isoelectric point (5.0). Kirkwood and Shumaker (1952a), on the other hand, pointed out that the charge fluctuations lead to a *fluctuating*

TABLE 6-8   Parameters Used to Evaluate Potential Functions

Parameter	Value
Albumin	
Molecular weight, M_p	69,000
Hydrated density, ρ	1.34 g/cm^2
Molecular volume, v_m	128,000 Å^2
Equivalent spherical radius, a	31.3 Å
Ellipsoid shape	
Semimajor axis, θ_a	70.5 Å
Semiminor axis, θ_b	20.8 Å
Dipole moment, μ	380×10^{-18} esu-cm
Polarizability, α_p	5950 Å^3
Characteristic frequency, ν_{0p}	3.06×10^{15} sec^{-1}
Charge number, Z	-20.4 at 7.4 pH
	-9.1 at 5.4 pH
	$+4.5$ at 4.5 pH
Root-mean-square charge number fluctuation, $\langle Z^2 \rangle^{1/2}$	3.5
Solvent (water)	
Polarizability, α_s	1.48
Characteristic frequency, ν_{0s}	4.35×10^{15} sec^{-1}
Dielectric constant	
Bulk value, ϵ	78.3
Local value at macroion surface, ϵ_s	4
Calculated	
Debye length, κ^{-1}	7.8 Å
Albumin surface potential, ψ_0	-23.5 mV at 7.4 pH
	-10.5 mV at 5.4 pH
	$+5.2$ mV at 4.5 pH
Hamaker constants	
A_p	7.27×10^{-13} ergs
A_s	5.25×10^{-3} ergs
A	1.65×10^{-14} ergs

Source: Data from Vilker et al., 1981.

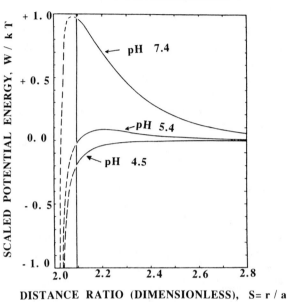

OSMOTIC PRESSURE OF PROTEIN SOLUTIONS

FIGURE 6-9.   Pair potential energy functions calculated for pH 4.5, 5.4, and 7.4 (according to Vilker et al., 1981).

dipole $\langle \mu \rangle_{av}$ that approaches the value of the permanent dipole moment, μ_0, of fixed charges (other than protons) at a pH value that is higher than the iso-electric point of the protein. In this case, $\langle \mu \rangle_{av}$ was predicted to be a "mono-tonically *increasing* function of pH." This prediction suggests, therefore, that the calculated curve for BSA solutions at pH 7.4 should have included the ap-propriate $\langle \mu \rangle_{av}$ value in the calculation of the interaction potential, in order to fit the data of Vilker et al. (1981) at this pH. It is unlikely that more than *three* virial coefficients would then be necessary to fit such data. *Preferential inter-actions*, however, should also be taken into account (Arakawa and Timasheff, 1982) in solutions containing small solutes as well.

HYDRATION OF PROTEINS, CARBOHYDRATES, AND STARCH

With some notable exceptions, water "bound" to food proteins, carbohydrates, and cereal starches is in *fast exchange* with bulk water. The hydration of such systems mostly involves *weak binding* of water to charged groups or hydrogen bonding to polar groups. The *molecular dynamics* of water in such systems are well-characterized by a *two-state, fast-exchange* model (Zimmerman and Brittin, 1957; Derbyshire, 1982), with the weakly bound water having average, fast correlation times on the order of 17 to 32 ps. In cereal starch the fast correlation time value determined by ^{17}O NMR relaxation increases to 30 ps at 70% corn starch (Yakubu, Baianu, and Orr, 1991).

In most proteins, including lysozyme, wheat gliadins, and corn zeins, bound water molecules also have a second type of slower motion associated with the tumbling of the proteins. These *slow correlation times* have values between 4.7 ns (for lysozyme) and 30 ns (for zeins), or longer for larger proteins (Halle et al., 1981; Kakalis and Baianu, 1988; Baianu et al., 1990). Water motion and hydration are strongly affected by protein aggregation. For example, water *trapping* occurs in lysozyme (Lioutas, Baianu, and Steinberg, 1987) and myofibrillar protein powders (Lioutas et al., 1988), and "water activity" decreases sharply upon water trapping. *Protein–protein interactions* modulate the two-"state," fast-exchange process that dominates the molecular dynamics of water in protein solutions (Kumosinski and Pessen, 1982). The binding of ions and protein aggregation can be quantitatively fitted by *nonlinear regression* analysis (Noggle, 1985; Motulsky and Ransnas, 1987) using the *thermodynamic linkage equations* first introduced by Wyman, which provide greater insight into the behavior of food proteins with salt than has been possible previously.

A unique behavior of water was found in potato starch, where water motions for a certain fraction are *anisotropic*, and the corresponding "tightly bound" water population is only in *slow exchange* with bulk water. The oriented water dominates the water sorption process in potato starch powders, and concentrated slurries (Yakubu, Baianu, and Orr, 1990). New technological applications could be developed on the basis of such observations.

Unique Hydration Properties of Potato Starch

Apart from its role as an energy nutrient, starch is gaining wide applications in the food industry as a dietary functional ingredient. Water plays a very important role in influencing the physical and functional properties of polysaccharides in food systems. The interactions of water with biopolymers have been studied successfully utilizing nuclear magnetic resonance (NMR) techniques that provide an improved understanding of the translational and rotational motions of water molecules in such systems. Despite the low magnetogyric ratio of the deuteron, and the relatively low sensitivity compared with proton NMR, deuterium (^2H) NMR (see Chapter 5 for details of the technique) is extensively used for studying *anisotropic* molecular motions, the dynamics of electron donor–acceptor complexes, protein hydration properties, gelatinization, retrogradation and conformational charges, heterocyclic bases, electrolyte solutions and starch structure.

Among the techniques used to study starch hydration properties, NMR provides a rapid, sensitive, and nondestructive means for investigating water interactions with starch over a wide range of hydration levels. Previous studies have utilized either pulsed-gradient ^{1}H NMR or NMR spectroscopy to study the molecular mobility and "state" of water in wheat starch, corn, potato, and wheat starches (continuous-wave ^{2}H NMR), waxy corn, pea, arrow root, tapioca, sweet potato starches, chemically modified waxy corn starch, amylopectin, maltodextrin, gelatinized corn starch, and corn syrups. Most of the work carried out by NMR has focused on the problem of identifying the different populations of water and their interactions with proteins and carbohydrates. Diffusion rates can also be measured by NMR in the presence of a field gradient (Stejskal and Tanner, 1965; Callaghan, Jolley, and Lelievre, 1979). (Further details of NMR applications are presented in Vol. II, especially in Chap. 9.)

Water Sorption Isotherms and Relative Vapor Pressure of Water in Foods

The knowledge of moisture sorption isotherms, usually in graphical form (Asbi and Baianu, 1986), has found applications in two major areas in food technology. Microbial growth (Kumosinski et al., 1988), enzymatic reactions, nonenzymatic browning, and lipid oxidations are some of the deteriorative mechanisms that are known to be dependent on water activity (Rockland and Beuchat, 1987; Rockland and Stewart, 1981). Such undesirable reactions would therefore be more rapid away from the optimum moisture content of the food (Matz, 1965) required for storage stability of dehydrated and intermediate moisture foods (as well as for packaging requirements and design).

The second major application of moisture sorption isotherms is in food engineering (Othmer and Sawyer, 1943), where the accurate computation of the equilibrium moisture-content values permits the optimization of drying as a food-unit operation, namely the calculations of drying times and efficiency in energy utilization (Van der Berg and Bruin, 1981).

Van der Berg and Bruin (1981) reviewed more than 70 model equations from the literature to mathematically describe *water sorption isotherms* found in foods, macromolecules, polymers, and so on. Among these are equations that were derived either from first principles of thermodynamics/statistical mechanics, or that were simply based on empirical or semiempirical considerations.

WATER ACTIVITY IN FOODS

Equilibria between Water Vapor and Bound Water

Consider a solid material binding water in equilibrium with water vapor in air at atmospheric pressure (standard conditions: T° = room temperature in absolute units (293 K); P° = vapor pressure of water at T°; $a_1^\circ = 1$; $f_1^\circ = f_1$ at P° and T°). Under these conditions the *activity of bound* water may be calculated as follows. In the vapor phase

$$\mu_1(\text{gas}) = \mu_1^\circ + \int_{P^\circ g}^1 \tilde{v}_1 \cdot dP = \mu_1^\circ + RT \ln f_1/f_1^\circ$$

In the bound water phase

$$\mu_1 = \mu_1^\circ + \int_{P^\circ}^1 \tilde{v}_1 \cdot dP = \mu_1^\circ + RT \ln s_{1B} \tag{6-48}$$

with the subscript B indicating the bound water population, or layer. At equilibrium (if this is indeed reached)

$$\ln f_1/f_1^\circ = \int_{P^\circ}^1 (RT)\tilde{v}_{1B}^1 \, dP + \ln (a_{1B})$$

Neglecting the integral, and assuming $f_1 = p_1$ and $f_1^\upsilon = p_1^\circ$, then $a_{1B} = p_1/p^\circ$ liquid. Note that equilibrium conditions may *not* always be present in foods.

Equilibrium between Liquid Bulk Water and Bound Water

Consider the isothermic transfer of a mole of water from liquid water to an adsorbent such as starch gel at a fixed moisture content. The standard states:

Liquid H_2O at $P = 1$ atm; $T = T^\circ$; $s_1^\circ = 1$
Bound H_2O at $P = 1$ atm; $T = T^\circ$, $a_{1B}^\circ = 1$

For liquid water

$$\mu_{1L} = \mu_1^\circ \tag{6-49}$$

where the subscript L stands for the *liquid* phase. For bound water

$$\mu_{1B} = \mu_1^\circ + RT \ln a_{1B} \tag{6-50}$$

Subtracting (6-49) from (6-50),

$$\Delta\mu_{1B} = RT \ln a_{1B}$$

or

$$\Delta\mu_{1B}/T = R \ln a_{1B} \tag{6-51}$$

Differentiating equation (6-51) one obtains

$$\partial(\mu_{1B}/T)/\partial T = R \cdot [\partial(\ln a_{1B})/\partial T]_{P,W} \tag{6-52}$$

From thermodynamics (Chap. 2), one has that

$$\Delta G = \Delta H - T\Delta S \quad \text{or} \quad \Delta\mu_{1B} = \Delta H_{1B} - T\Delta S_{1B}$$

and

$$\Delta\mu_{1B}/T = \Delta H_{1B}/T - \Delta S_{1B} \tag{6-53}$$

Differentiating, equation (6-53), one has

$$\partial(\mu_{1B}/T)/\partial T = (1/T) \cdot \partial(\Delta H_{1B})/\partial T - \Delta H_{1B}/T^2 - \partial(S_{1B})/\partial T \tag{6-54}$$

For constant pressure, one has

$$\Delta H_1 c_p = \partial\Delta H_1/\partial T = T(\partial\Delta S_1/\partial T) \tag{6-55}$$

Therefore,

$$\partial(\Delta\mu_{1B})/\partial T = -\Delta H_{1B}/T^2 \tag{6-56}$$

or equating (6-52) and (6-56), and rearranging the terms, one obtains

$$[\partial(\ln a_{1B})/\partial T]_{P,W} = -\Delta H_{1B}/RT^2 \tag{6-57}$$

From equation (6-49) one obtains

$$d(\ln p_1/p_1^\circ) = -\Delta H_{1B} \, dT/RT^2 \tag{6-58}$$

or assuming that H_{1B} is constant, then

$$d(\ln p_1/p_1^\circ) = \Delta H_{1B} \cdot (1/RT) + \text{constant} \tag{6-59}$$

For a liquid in equilibrium with its vapor, it can be demonstrated (assuming $f = p$) that

$$\ln p_1^\circ = -(\Delta H_v/R) \cdot 1/T + C'$$

$$1/T = -R \ln p_1^\circ/\Delta H_v + RC'/\Delta H_v \tag{6-60}$$

Substituting equation (6-29) in (6-27)

$$\ln p_1/p_1^\circ = \Delta H_{1B}/R - R \ln p_1^\circ/\Delta H_v + RC'/\Delta H_v$$

or

$$\ln p_1 = (-\Delta H_{1B}/\Delta H_v + 1) \cdot \ln p_1^\circ + (\Delta H_{1B}C'/\Delta H_v + C) \quad (6\text{-}61)$$

Therefore, a plot of the logarithm of the vapor pressure over the adsorbent at constant moisture content versus the logarithm of the vapor pressure over pure water at the same temperature will be a straight line with a slope equal to $(\Delta H_v - \Delta H_{1B})/\Delta H_v$.

Determination of Specific Surface Coefficients

If the Langmuir adsorption isotherm (Fig. 6-10) is rewritten in the form

$$S^1 = B^1 \cdot k \cdot p/(1 + k \cdot p) \quad (6\text{-}62)$$

where S^1 is the number of molecules of adsorbate per gram, and B^1 is the number of sites per gram, and if the adsorbate covers the entire surface of the adsorbent in a monolayer (Brunauer, Emmett, and Teller, 1938), then the specific surface is

$$S^1 = B^1 \cdot A \quad (6\text{-}63)$$

where A is the cross-sectional area of an adsorbate molecule in cm^2. Also, if several different kinds of sites exist on the adsorbent, each will have its own B

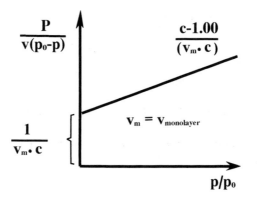

FIGURE 6-10.   Linear plot of the Langmuir isotherm for adsorption of gases in solids.

and k, and therefore,

$$S = \sum_i (B_i/N) \cdot k_i \cdot P/(1 + k_i \cdot P) \qquad (6\text{-}64)$$

Capillary Condensation Theory

Kistler, Fisher, and Freeman observed that in a particulate material with pores, condensation could take place in the capillary pores if the gas was at a temperature below its critical point, according to an application of the Boltzmann equation:

$$P/P^\circ = e^{-MA\gamma/WRT} \qquad (6\text{-}65)$$

where P is the pressure of the gas, P° is vapor pressure over liquified gas at temperature T, M is the molecular weight of gas, A is the area of one gram of adsorbent, γ is the surface tension of liquified gas, R is the universal gas constant, W is the weight of adsorbate per gram of adsorbent, and $A\gamma/W$ is the work required to pump out one gram of adsorbent. This equation neglects monomolecular adsorption, assumes that the capillaries are either full or empty, and assumes *uniform* capillaries. If the structure is nonuniform, however, this equation becomes

$$P/P^\circ = e^{(-M\gamma/RT \cdot dA/dW)} \qquad (6\text{-}66)$$

or

$$dA/dW = 2.303 \cdot RT/M\gamma \cdot (\log P^\circ/P) \qquad (6\text{-}67)$$

Adsorption of H₂O on SiO₂ Aerogel at 35.4°C

The area under curve II is $\int_0^{P^\circ} [\log P^\circ/P] \cdot dw$. Therefore, the capillary surface area is

$$A = 2.303 \cdot RT/M\gamma \cdot \left(\int_0^{P^\circ} [\log P^\circ/P] \cdot dw \right) \qquad (6\text{-}68)$$

if corrected for the monolayer absorption.

Monolayer and Multimolecular Layer Theories

By analogy with the Langmuir derivation for a monomolecular layer, one can consider that $s_0, s_1, s_2 \cdot \cdot \cdot s_i \cdot \cdot \cdot$ represent the surface areas covered by the

0, 1, 2 · · · *i* · · · *layers* of adsorbed molecules (Brunauer, Emmett, and Teller, 1938; Brunauer, 1945). At equilibrium, the rate of condensation on the *bare* surface must be equal to the rate of evaporation from the first layer (neglecting edge effects), and s_0 is a constant, or

$$a_1 p s_0 = b_1 s_1 e^{-H_1/RT} \tag{6-69}$$

where p is the pressure of the gas, H_1 is the heat of adsorption of the first layer, and a_1 and b_1 are constants characteristic of the materials.

At equilibrium, s_1 must also be constant; therefore:

$$a_2 p s_1 + b_1 s_1 e^{-H_1/RT} = b_2 s_2 e^{-H_2/RT} + a_1 p s_0 \tag{6-70}$$

From equations (6-69) and (6-70)

$$a_2 p s_1 + b_2 s_2 e^{-H_2/RT} \tag{6-71}$$

Extending this argument to the i layers, one obtains,

$$a_3 p s_2 + b_3 s_3 e^{-H_3/RT}$$
$$\vdots \qquad \qquad \vdots$$
$$a_i p s_{i-1} = b_i s_i e^{-H/RT} \tag{6-72}$$

The total area of the surface is

$$A = \sum_{i=0}^{\infty} (S_i) \tag{6-73}$$

and the volume of gas adsorbed is

$$v = v_0 \cdot \sum_{i=0}^{\infty} (i S_i) \tag{6-74}$$

where v_0 is the volume adsorbed per square centimeter when the surface is completely covered by a *monolayer* (Brunauer, Emmett, and Teller (BET) theory). Therefore, one obtains from equations (6-73) and (6-74)

$$v/A v_0 = v/v_m = v_0 \cdot \sum_{i=0}^{\infty} (iS) \Big/ \sum_{i=0}^{\infty} (S) \tag{6-75}$$

where v_m is the volume adsorbed when the surface is completely covered by a *monolayer*. Assuming uniform sites, one has that

$$\Delta H = \Delta H_1 = \cdots = \Delta H_i = \Delta H_L \qquad (6\text{-}76)$$

where ΔH_L is the heat of liquefaction of the gas, and

$$b_2/a_2 = \Delta b_3/a_3 = \cdots b_i/a_i = g \qquad (6\text{-}77)$$

This is the same as stating that the second and higher layer properties are the same as in the liquid state. Expressing $s_1, s_2, \cdots, s_i$ in terms of s_0

$$s_1 = y s_0 \qquad \text{where} \quad y = a_1/b_1 \cdot p e^{(\Delta H_1/RT)} \qquad (6\text{-}78)$$

$$s_2 = x s_1 \qquad \text{where} \quad x = (p/g) \cdot e^{(\Delta H_1/RT)} \qquad (6\text{-}79)$$

$$s_3 = x s_2 = x^2 s_1$$

$$s_i = x s_{i-1} x^{i-1} s_1 = y x s_0 = c x^i s_0 \qquad (6\text{-}80)$$

where

$$c = y/x = a_1/b_1 \cdot g \cdot e^{(\Delta H_1 - \Delta H_L)RT}. \qquad (6\text{-}81)$$

Substituting equation (6-60) from (6-54)

$$v/v_m = c s_0 \cdot v_0 \cdot \sum_{i=0}^{\infty} (i + x^i)/S_0 \left[1 + c \cdot \sum_{i=0}^{\infty} (x^i) \right] \qquad (6\text{-}82)$$

The sum in the denominator is the sum of an infinite geometric progression

$$\sum_{i=0}^{\infty} (x^i) = x/(1 - x) \qquad (6\text{-}83)$$

The sum in the numerator is

$$\sum_{i=0}^{\infty} (i \cdot x^i) = \frac{x d}{dx} \left[\sum_{i=0}^{\infty} (x^i) \right] = \frac{x}{(1 - x)^2} \qquad (6\text{-}84)$$

Therefore, one obtains the relatively simple equation

$$\frac{v}{v_m} = \frac{cx}{(1 - x)(1 - x + cx)} \qquad (6\text{-}85)$$

At the saturation pressure of the gas, p_0, for a free surface, x must be equal to 1.00 to make $v = \infty$ at $p = p_0$. Therefore,

$$\frac{p_0}{g} \cdot e^{\Delta H_L/RT} = 1 \quad \text{and} \quad x = \frac{p}{p_0} \tag{6-86}$$

Substituting in equation (6-65), one obtains the following isotherm

$$v = v_s \cdot c \cdot p/(p_0 - p)[1 + (c - 1)p/p_0] \tag{6-87}$$

For nonuniform sites in *heterogeneous* systems, D'Arcy and Watt (1970) proposed a different *multilayer* theory.

Evaluation of c and v_m

Rearranging equation (6-86), one obtains

$$p/v(p_0 - p) = 1/V_m c + (c - 1)/v_m c \cdot p/p_0 \tag{6-88}$$

A plot of $p/v(p_0 - p)$ against p/p_0 will yield a straight line. If the thickness of the adsorbed layer is limited either by the dimensions of the pores or for any other reason to n layers, it can be shown that the isotherm takes the form

$$v = v_m \cdot cx[1 - (n + 1)x^n + nx^{n+1}]/(1 - x)[1 + (c - 1)x - cx^{n+1}], \tag{6-89}$$

where $x = p/p_0$ is the relative vapor pressure

Type I Isotherms

When $n = 1$, equation (6-68) reduces to the *Langmuir adsorption* isotherm (Fig. 6-10) (ethyl chloride on charcoal at 0°C)

$$v = v_m \cdot c \cdot x/(1 + cx) \tag{6-90}$$

or

$$v = v_m \cdot c \cdot p/p/(1 + c \cdot p/p_0) \tag{6-91}$$

This equation uses data gathered at one temperature to provide information of the adsorption phenomena at other temperatures, since

$$c = (a_1/b_1) \cdot g \cdot e^{(\Delta H_1 - \Delta H_L)/RT} = e^{(\Delta H_1 - \Delta H_L)/RT} \tag{6-92}$$

Experimentally the equation fits other temperatures well at $p/p_0 > 0.1$.

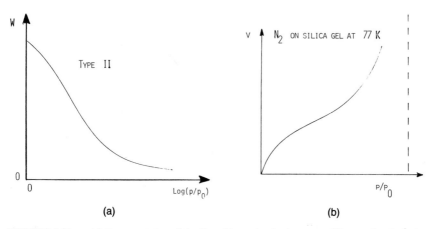

FIGURE 6-11. (a) Representation of the Type II sorption isotherm as a W versus log (p/p_0), where p is the pressure of the adsorbed gas and p_0 is the normal pressure of the free, unadsorbed gas. (b) Type II adsorption isotherm of N_2 gas on silica gel near 77 K, represented as the volume of adsorbed gas against the relative vapor pressure of the adsorbed N_2. A wide range of foods and biopolymers exhibit Type II isotherms.

Type II Isotherms
When $n = \infty$, type II isotherms (Fig. 6-11) are obtained

$$v = v \cdot c \cdot p/(p_0 - p) \cdot [1 + (c - 1) \cdot p/p_0] \qquad (6\text{-}93)$$

Adsorption isotherms calculated at temperatures other than the one determined agree well at $p/p_0 > 0.05$. This isotherm type is important in foods and biopolymers.

Type III Isotherms
When $H_1 = H_L$ or $H_1 < H_L$ and $n = \infty$, type III isotherms are obtained (Fig. 6-12).

Type IV Isotherms
A modified BET theory has to consider finite capillaries consisting of two plane-parallel walls, n layers apart. The last molecule to fill a layer then has a heat of adsorption equal to $(AH_L + Q)$, since two interfaces are involved. In a way similar to the derivation of equations (6-76) represented in Figure 6-13

$$v = v_{m/2} \cdot x \cdot dA/(A \cdot dx) \qquad (6\text{-}94)$$

where

$$A = s_\infty[1 + 2cxn] = c^2 \cdot x^2 \cdot dn/dx + (n \cdot c^2 - c^2 + 2 \cdot c)q \cdot x^n$$
$$(6\text{-}95)$$

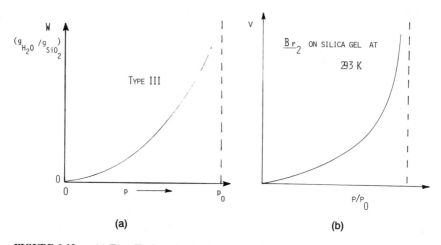

FIGURE 6-12.   (*a*) Type II adsorption isotherm for water adsorbed in porous glass. (*b*) Type III adsorption isotherm of Br_2 on silica gel at 293 K.

and

$$n = (1 - x^{n-1})/(1 - x) \qquad (6\text{-}96)$$

and also,

$$q = e \cdot (Q/RT) \qquad (6\text{-}97)$$

When $c \gg 1$, this equation fits a type IV isotherm quite satisfactorily, though not perfectly, because of the following assumptions in the derivation:

1. Capillaries are of uniform size
2. Walls of capillaries are parallel
3. Capillaries are open at sides

Type V Isotherms
When $C < 1$, equation (6-94) fits a type V isotherm quite satisfactorily, though not perfectly (Fig. 6-14).

Hysteresis
Equation (6-94) does not fit the hysteresis phenomena frequently encountered in types IV and V isotherms (Fig. 6-15). At present, the problem of hysteresis

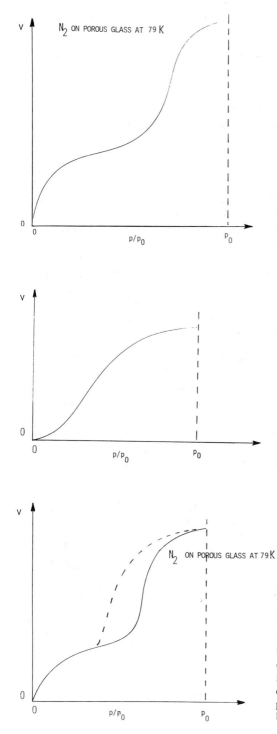

FIGURE 6-13. Adsorption isotherm of N_2 on porous glass at 79 K (Type IV isotherm).

FIGURE 6-14. Type V adsorption isotherm.

FIGURE 6-15. Hysteresis effect on the adsorption–desorption cycle for N_2 on porous glass at 79 K, indicating perhaps the effects of gas- or liquid-filled capillaries on the subsequent adsorption–desorption processes. The interrupted curve represents the desorption measurements, presumably after equilibration has been reached.

253

has not been completely solved (see also Franks, 1982, 1991). An example of an application to dried whole milk follows:

1. N_2 and He gases are used.
2. The spray dried milk surface area is determined to be 9200 cm^2/g.
3. Instant and foam dried milk have a surface area of 2400 to 5600 cm^2/g.

References

Arakawa, T., and S. N. Timasheff. 1982. Preferential interactions of proteins with salts in concentrated solutions. *Biochemistry* **21**:6543–6552.

Asbi, B. A., and I. C. Baianu, 1986. An equation for fitting moisture sorption isotherms of food proteins. *J. Agric. Food Chem.* **34**:494–496.

Baianu, I. C., T. F. Kumosinski, P. J. Bechtel, P. A. Myers-Betts, L. Kakalis, A. Mora-Guttierrez, and P. Yakubu. 1990. Multinuclear spin relaxation and high-resolution nuclear magnetic resonance studies of food proteins, agriculturally important materials and related systems. In *NMR Applications in Biopolymers*, J. W. Finley, et al., ed., 361–389. New York: Plenum.

Ben-Naim, A., and F. H. Stillinger, Jr. 1972. In *Water and Aqueous Solutions*, R. A. Horne, ed., 295–330. New York: Wiley-Interscience.

Brunauer, S. 1945. *The Adsorption of Gases and Vapors*, vol. 1. Princeton, N.J.: Princeton University Press.

Brunauer, S., P. H. Emmett, and E. Teller, 1938. Adsorption of gases in multimolecular layers. *J. Am. Chem. Soc.* **60**:309–316.

Callaghan, P. T., K. W. Jolley, and J. Lelievre. 1979. Diffusion of water in the endosperm tissue of wheat grains as studied by pulsed field gradient nuclear magnetic resonance. *Biophys. J.* **28**:133–141.

Chapman, D. L. 1913. A contribution to the theory of electrocapillarity. *Phil. Mag.* **25**:475.

Conway, B. E. 1981. *Ionic Hydration in Chemistry and Biophysics.* Amsterdam, The Netherlands: Elsevier.

D'Arcy, R. L., and I. C. Watt. 1970. Analysis of sorption isotherms of non-homogeneous sorbents. *Trans. Faraday Soc.* **66**:1236–1245.

Derbyshire, W. 1982. The dynamics of water in heterogeneous systems with emphasis on subzero temperatures. In *Water: A Comprehensive Treatise*, vol. 7, F. Franks, ed., 339–469. New York: Plenum.

Emmett, P. H. 1942. Adsorption of gases in multimolecular layers. *Adv. Colloid Sci.* **1:** 1–36.

Franks, F. 1982. Water activity as a measure of biological viability and quality control. *Cereal Foods World* **27**:403–407.

Franks, F. 1991. Water activity: A credible measure of food safety and quality? *Trends Food Sci. Technol.* March, 68–72.

Gouy, G. 1910. *J. de Physique* **9**:457–462.

Grahame, D. C., and R. Parsons. 1961. Components of charge and potential in the inner region of the electrical double layer. *J. Am. Chem. Soc.* **83**:1291–1296.

Halle, B., T. Andersson, S. Forsen, and B. Lindman, 1981. Protein hydration from water oxygen-17 magnetic relaxation. *J. Am. Chem. Soc.* **103**:500–508.

Hamaker, H. C. 1937. *Physica* **4**:1058–1067.

Kakalis, L. T., and I. C. Baianu. 1988. Oxygen-17 and deuterium nuclear magnetic resonance relaxation studies in solution: Field dispersion, concentration, pH/pD and protein activity dependence. *Arch. Biochem. Biophys.* **267**:829–841.

Keesom, W. H. 1921. Die van der Waalsschen Kokätionskräfte. *Phys. Z.* **22**:129–141, 643–644.

Kirkwood, J. G. 1932. Quantenmechanische Berechnung der Konstanten einiger polaren Molekule. *Phys. Z.* **33**:259–265.

Kirkwood, J. G. 1934. Theory of solutions of molecules containing widely separated charges with special application to zwitterions. *J. Chem. Phys.* **2**:351–361.

Kirkwood, J. G. 1939. Dielectric polarization of polar liquids. *J. Chem. Phys.* **7**:911–919.

Kirkwood, J. G., and J. B. Shumaker. 1952a. The influence of dipole moment fluctuations on the dielectric increment of proteins in solution. *Proc. Natl. Acad. Sci. U.S.A.* **38**:855–862.

Kirkwood, J. G., and J. B. Shumaker. 1952b. Forces between protein molecules in solution arising from fluctuations in proton charge and configuration. *Proc. Natl. Acad. Sci. U.S.A.* **38**:863–871.

Kruyt, H. R. 1952. *Coloid Science*, vol. 1. New York: Elsevier.

Kumosinski, T. F., and H. Pessen. 1982. A deuteron and proton magnetic resonance relaxation study of β-lactoglobulin A association: Some approaches to the Scatchard hydration of globular proteins. *Arch. Biochem. Biophys.* **218**:286–302.

Kumosinski, T. F., H. Pessen, and H. M. Farrell, Jr. 1988. Proteins as potential bacterial growth inhibitors in foods: Suppression of the activity of water by proteins as determined by NMR relaxation methods. *J. Ind. Microbiol.* **3**:147–155.

Linderstrøm-Lang, K. 1924. *C. R. Trav. Lab. Carlsberg.* **15**:7.

Lioutas, T. S., I. C. Baianu, and M. P. Steinberg. 1987. Sorption equilibrium and hydration studies of lysozyme: Water activity and 360 MHz proton NMR measurements. *J. Agric. Food Chem.* **35**:133–137.

Lioutas, T. S., I. C. Baianu, P. J. Bechtel, and M. P. Steinberg, 1988. Oxygen-17 and Sodium-23 nuclear magnetic resonance studies of myofibrillar protein interactions with water and electrolytes in relation to sorption isotherms. *J. Agric. Food Chem.* **36**:437–444.

Matz, A. 1965. *Water in Foods.* Westport, Conn.: AVI.

Momany, F. A., R. F. McGuire, A. W. Burgers, and H. A. Scheraga. 1975. Energy parameters in polypeptides. VII. Geometric parameters, partial atomic charges, nonbonded interactions, and intrinsic torsional potential for the naturally occurring amino acids. *J. Phys. Chem.* **79**:2361–2381.

Motulsky, H. J., and L. A. Ransnas. 1987. Fitting curves to data using nonlinear regression: A practical and nonmathematical review. *Fed. Am. Soc. Exp. Biol.* **1**:365–374.

Narten, A. H., M. D. Danford, and A. H. Levy, 1967. X-ray diffraction studies of liquid water at various temperatures. *Faraday Soc. Disc.* **34**:7357–7366.

Noggel, J. H. 1985. *Physical Chemistry on a Microcomputer*, 145–165. Boston: Little, Brown.

Othmer, D. F., and F. G. Sawyer. 1943. Correlating adsorption data: Temperature, pressure, concentration heat. *Ind. Eng. Chem.* **35**:1269–1276.

Parker, S., ed. 1983. *McGraw-Hill Encyclopedia of Chemistry.* New York: McGraw-Hill.

Rockland, L. B., and L. R. Beuchat, eds. 1987. *Water Activity, Theory and Applications to Foods*. New York: Marcel Dekker.

Rockland, L. B., and G. F. Stewart, eds. 1981. *Water Activity: Influences on Food Quality*. New York: Academic Press.

Scatchard, G. 1943. In *Proteins, Amino Acids, and Peptides*, R. Cohn and J. Edsall, eds. New York: Reinhold.

Scatchard, G., and J. G. Kirkwood, 1932. *Phys. Z.* **33**:297.

Schachman, H. K. 1959. *Ultracentrifugation in Biochemistry*, 226. New York: Academic Press.

Schick, M. J. 1987. *Nonionic Surfactants Physical Chemistry*. 1003. New York and Basel: Marcel Dekker.

Schulz, G. E., and R. H. Schirmer. 1979. *Principles of Protein Structure*. New York, Heidelberg, Berlin: Springer-Verlag.

Slater, J. C., and J. G. Kirkwood, 1931. The van der Waals forces in gases. *Phys. Rev.* **37**:682–697.

Soper, A. K., and M. G. Phillips. 1986. A new determination of the structure of water at 25°C. *Chem. Phys.* **107**:47–60.

Stejskal, E. O., and J. E. Tanner. 1965. Spin diffusion measurements: Spin echoes in the presence of a time-dependent field gradient. *J. Chem. Phys.* **42**:288–292.

Stern, O. 1924. Zur Theorie der Elektrolytischen Doppelschicht. *Z. Elektrochem.* **30**:508–516.

Stokes, R. H., and R. A. Robinson, 1948. Ionic hydration activity in electrolyte solutions. *J. Am. Chem. Soc.* **70**:1870–1878.

Stokes, R. H., and R. A. Robinson. 1949. *Ann. N.Y. Acad. Sci.* **51**:593.

Tanford, C. 1957. The location of electrostatic charges in Kirkwood's model of organic ions. *J. Am. Chem. Soc.* **79**:5348–5352.

Tanford, C. 1961. *Physical Chemistry of Macromolecules*, 227, 293, 352, 563. New York: Wiley.

Timasheff, S. N., and B. D. Coleman. 1960. On light-scattering studies of isoionic proteins. *Arch. Biochem. Biophys.* **87**:63–69.

Timasheff, S. N., and M. J. Kronman. 1959. The extrapolation of light scattering data to zero concentration. *Arch. Biochem. Biophys.* **83**:60–75.

Timasheff, S. N., H. M. Dintzis, J. G. Kirkwood, and B. D. Coleman. 1957. Light scattering investigations of charge fluctuations in isoionic serum albumin solutions. *J. Am. Chem. Soc.* **79**:782–791.

Van den Berg, C., and S. Bruin. 1981. Water activity and estimation in food systems. In *Water Activity: Influences on Food Quality*, L. B. Rockland and G. F. Stewart, eds. New York: Academic Press.

Vilker, V. L., et al. 1981. Osmotic pressure studies of concentrated serum albumin solutions. *J. Colloid Interface Sci.* **79**(2):548–566.

Yakubu, P., I. C. Baianu, and P. H. Orr. 1991. Unique hydration behavior of potato starch as determined by deuterium nuclear magnetic resonance. *J. Food Sci.* **55**(2):458–461.

Zimmerman, J. R., and W. E. Brittin. 1957. Nuclear magnetic resonance studies in multiple phase systems: Lifetime of a water molecule in an adsorbing phase on silica. *J. Phys. Chem.* **61**:1328–1333.

7

Physical Chemistry in the Food Processing Pilot Plant

Michael F. Kozempel and Peggy Tomasula

NOMENCLATURE

c coefficient
C solute concentration in the potato
C_1 initial solute concentration in the potato
C_e equilibrium solute concentration within the potato
C_p heat capacity
D effective diffusion coefficient, reactor diameter
$dP/d\theta$ flow rate of mash
F fraction of tracer fluid in exit stream
h convective heat transfer coefficient to air
I internal age distribution
L reactor length, drum width, nominal thickness of cut pieces
m molality, number of terms in equation (7-4)
M moisture content of potato
P potato flow rate
R gas constant, radius of drum dryer

RI refractive index of solution
RI_0 refractive index of initial solution
S solute concentration in blanch water
S_1 solute concentration in entering blanch water
T temperature
t time
T_a ambient or dry-bulb temperature
T_f equilibrium mash temperature
T_s steam temperature
T_w wet-bulb temperature
V volume of the blanch water
W water flow rate
x diffusion path
Δr mash thickness
δH differential heat of solution
δx small distance of travel on drum surface
θ reduced residence time
λ latent heat of vaporization
ρ density of the blanch water

τ residence time of the hot- ω drum speed, rpm
    water blancher

The principles of physical chemistry apply to both pilot plant or commercial processing as well as to the laboratory. The problem is identifying the significant principles that control processes from the myriad of confounding, nonsignificant parameters. It is only with the advent of the computer that it is possible to handle a sufficient number of variables to properly account for some of the principles involved. The confounding parameters may be referred to as *nonidealities*. For example, deviations from laboratory results may appear because flow patterns in large pilot plant or commercial-scale reactors are rarely simple (perfectly mixed, or perfectly plug, flow). The actual flow pattern may cause difficult temperature control or lower reaction yields than expected for a simple flow. In addition, food processing systems are *rarely homogeneous*. These systems are usually of the following varieties: gas–liquid, as in evaporative concentration; gas–solid, as in convection drying; or, liquid–solid, as in a leaching process. Catalytic components (i.e., enzymes) are usually unstable under processing conditions. It is the determination of the controlling physical chemical principles to which much of pilot plant research is addressed. The application of physical chemistry in the pilot plant usually involves developing *mathematical models*. Using the appropriate physical chemical laws as a basis, the models aid the engineer in achieving a greater understanding of a process and permit extrapolation to processing conditions for which many of the controlling variables cannot be readily measured experimentally in the laboratory.

Consider the drying of a bed of potato slices in a commercial tunnel dryer. In this process, internal moisture from the drying potatoes diffuses to the surface, where it evaporates. Because of the variability in the raw material (potatoes), each slice is a different size, shape, and composition; there are air gaps between the slices; hence, air flow and contact is nonuniform. There is enzymatic and nonenzymatic browning. If the bed is fairly uniform, it is possible to determine an appropriate drying mechanism and to calculate the drying rate for an existing dryer from parameters dependent on measured process conditions. To design a new dryer, mathematical models for drying must include mass and energy balances for the dryer and the slices. The main variables are the temperature and moisture content of the potato slices leaving the dryer plus the drying time and temperature of the dryer.

Physical property data such as heat and mass transfer coefficients, heat capacities, thermal diffusivities, and shrinkage data are required in the drying of fruits and vegetables. Although some data are available in the literature (Singh, 1982; Balaban, 1988), it must usually be estimated because the variability between individual pieces due to cultural practices, cultivar, maturity, soil type, area where grown, environment, and storage conditions after harvest, makes it time-consuming and costly to compile data. Correlations based on generalized methods are often used (Rotstein, 1990). The real examples that follow give a better understanding of the application of physical chemistry in pilot plant research.

MIXING

An overriding phenomenon associated with almost all pilot plant processing is concerned with flow patterns during mixing and mass transport. Mass and energy transfer are dependent to a great extent on *mixing*. The principles governing mass and energy transfer are based on the assumption of perfect mixing, something that can only be achieved in a practical sense in the laboratory and approached in the pilot plant. Therefore, it is necessary to determine the actual mixing or flow pattern to develop a model for the real situation.

An excellent treatment of the theory of *residence time distribution* (RTD) is presented in Levenspiel (1972). It is assumed that actual flow patterns range between the ideal states of backmixing and plug flow, and may be modeled by a combination of backmixing and plug flow elements with dead regions and bypassing.

Backmixing is complete or perfect mixing, in which flow is turbulent and $L/D = 1$, where L is vessel length and D is vessel diameter. All fluid elements in the vessel have the same intrinsic properties, for example, concentration, temperature, and dwell time.

Plug flow is normally achieved in pipes under fairly high flow rates with Reynolds numbers greater than 10,000 and L/D ratios greater than 50. There is no axial mixing. It may be thought of as perfect radial mixing in a pipe or a series of infinitely small backmix reactors.

"Dead water" is thought of as fluid that moves slower relative to the fluid bulk or as a stagnant region in a vessel. It is caused by regions in the vessel that are virtually *unmixed*. Bypassing results when some of the fluid moves faster than the bulk of the fluid. Here the fluid does not mix (opposite of dead water), but simply passes through the vessel with no interaction with the bulk of the fluid. In the pilot plant none of these extremes would normally be encountered. All mixing would be a combination of two or more of these mixing patterns.

Generally, there are two problems associated with characterizing fluid mixing in a vessel. How is the RTD determined, and how are the results applied mathematically to the process under study? (In some cases, it may be necessary to treat the micromixing patterns, but normally characterizing macromixing is sufficient.) A simple way to experimentally measure RTD is with a *tracer* that is naturally part of or similar to the fluid component of interest.

For example, in developing a process for a cocoa butter substitute, it was necessary to determine the RTD of the fractional crystallization system. Fat was used as the tracer. The RTD determination was started by establishing a steady-state flow of fat and solvent through the crystallizing system and maintaining temperature sufficiently high to avoid crystallization. At steady state there was no change in concentration, as monitored by the refractive index (RI). At zero time, the fat feed was replaced with a solvent (fat-free) feed, and the effluent refractive index (concentration of fat) was recorded for twice the nominal residence time. Dividing the RI by RI_0 gives F, the *fraction of tracer fluid* in the exit stream, with I equal to $1 - F$:

$$F = \frac{RI}{RI_0} \qquad (7\text{-}1)$$

$$I = 1 - F \qquad (7\text{-}2)$$

The resulting data were plotted as I, the internal age distribution function, versus the *reduced residence time*, θ, as shown in Figure 7-1. The figure indicates the type of flow pattern, assuming the actual mixing can be represented as a mix of two ideal flow patterns. In the cocoa butter substitute study the flow pattern was best represented by the combination *backmixing* and *plug flow*. The

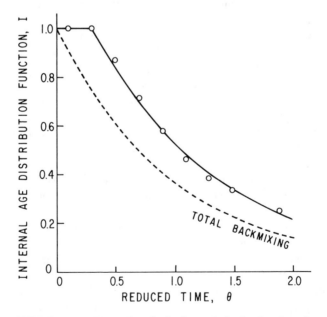

FIGURE 7-1.   Residence time distribution study for fractionating tallow.

horizontal line at $I = 1$ is plug flow, and the exponential decay curve corresponds to backmixing. Using one of the simple flow models found in Levenspiel (1972) to describe the curve, indicated that there was 74% backmixing and 26% plug flow.

This was desirable for this particular process. The crystallization step was very quick, about 10 minutes. A large source of nuclei were needed to produce the proper crystals so quickly. Plug flow would contain no nuclei, and the process would be expected to give unsuitable crystals due to spontaneous nucleation. With this large amount of backmixing there were sufficient nuclei present. In scale-up, backmixing would be a critical parameter to keep constant.

In a study of hot-water blanching of potatoes, a tracer was added to the feed stream, as opposed to the previous example in which the drop in tracer concentration was followed after tracer addition was terminated. To study the flow patterns in a hot-water blancher for blanching, a steady-state blancher operation was established with plain water, that is, the blancher was operated as if potatoes were being blanched, even though no potatoes were fed into it. At time equal zero, a lactose solution was fed in, and its concentration was followed at the exit of the blancher. Again, RI was a convenient way to follow the tracer. The flow pattern closely matched the same model—a mixed model of backmixing and plug flow.

Another method of applying the results of the RTD mathematically to the process under study is by treating the fluid stream as the sum of ideal streams, for example, backmixed, plug flow, dead water, or by-pass. The ideal streams are treated separately and then combined. An example of this method is described in the next section.

DIFFUSION

In potato processing, potato pieces go through a hot-water bath or blancher to leach sugars (especially reducing sugars, such as glucose) that contributes to nonenzymatic browning (Talburt and Smith, 1987). Sugars diffuse from the interior of the potato pieces to the surface, where they dissolve in the hot water. However, the rate-controlling step could be mass transfer at the surface, due to resistance at the boundary layer, or *diffusion* through the piece. What controls the rate of leaching, and how is the rate-controlling step and its mechanism determined?

A diagram of the pilot plant equipment is shown in Figure 7-2. Water was circulated through a heat exchanger to the blancher and back, as shown. The potatoes were passed through the blancher in small baskets, the basket rate of travel through the blancher simulating potato flow rate. The concentration of glucose and potassium in the exit water was monitored.

Because of the high recycle rate this system closely resembled an ideal *con-*

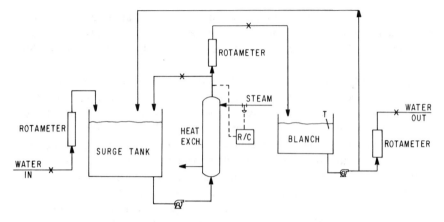

FIGURE 7-2.   Process for simulating blanching potatoes.

tinuous-flow stirred-tank reactor (CSTR), or backmix reactor, with complete backmixing and uniform temperature. Although glucose was the component of interest, potassium concentration was used as a flow tracer. Glucose in the water samples is subject to bacterial degradation, but potassium can be readily analyzed even if all potato solids were to decompose. Typical experimental data are shown in Figure 7-3.

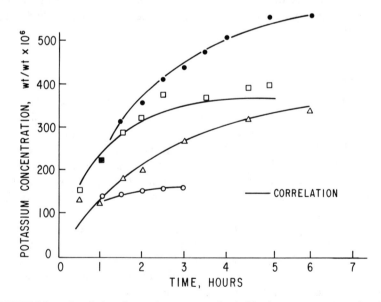

FIGURE 7-3.   Correlation of potassium concentration in blancher water vs. processing time.

The experimental data were best represented by a *diffusion model*. Mass transfer at the surface was found to be insignificant. Diffusion in potatoes may be represented by *Fick's second law:*

$$\frac{\delta C}{\delta t} = D \frac{\delta^2 C}{\delta x^2} \tag{7-3}$$

A. B. Newman (1931) presented a simple series expansion to solve the partial differential equation of Fick's law for a French fry strip of potato:

$$C = C_e + (C_1 - C_e) \frac{8}{\pi^2} \sum_{m=0}^{\infty} \frac{1}{(2m + 1)^2} \exp\left[-D(2m + 1)^2 \pi^2 \frac{\tau}{L^2} \right] \tag{7-4}$$

Use of this model assumes that the potato consists of an *insoluble* and *isotropic matrix of starch*, the *major solid component of the potato*, and other solid components (such as pectin), and a liquid component—water. Equating C_e to the solute concentration in the water, S, and making a mass rate balance over the hot-water blancher (Kozempel et al., 1981a), the predicted or theoretical values of potassium concentration in the blanch water (also other diffusing components such as glucose) can be calculated from

$$PMC_1 \left[1 - \frac{8}{\pi^2} \exp\left(-\frac{\pi^2 D\tau}{L^2} \right) \right] - PMS \left[1 - \frac{8}{\pi^2} \exp\left(-\frac{\pi^2 D\tau}{L^2} \right) \right]$$

$$+ W(S_1 - S)$$

$$= \left[V\rho + \frac{PM\tau}{2} \left(1 - \frac{8}{\pi^2} \exp\left(-\frac{\pi^2 D\tau}{L^2} \right) \right) \right] \frac{dS}{d\theta} \tag{7-5}$$

Assuming uniform temperature, piece size, and complete backmixing the value of the *effective diffusion coefficient*, D, may be calculated from equation (7-5). The D is termed an effective diffusion coefficient because the potato is not truly isotropic; its value is influenced by other diffusing components of the potato and the diffusional path. To determine the value of D from the data of Figure 7-3, a pattern optimization program was used. The correlation coefficients for the curves in Figure 7-3 were 0.9 or above.

Having established diffusion as the controlling step for leaching during blanching, the research went in two directions. At the bench level, Tomasula and Kozempel (1989) determined the variation of the diffusion coefficients of glucose, potassium, and magnesium in potatoes with temperature. At the bench, the effect of flow on leaching can be controlled to get "true" effective diffusivities. The data showed a change in slope at the gelatinization temperature (60°–

65°C). Effective diffusion coefficients were correlated with reciprocal temperature over three temperature ranges using an Arrhenius-type equation over the full temperature range (45°–90°C) and a separate polynomial with temperature-dependent coefficients for the ranges (45°–60°C and 65°–90°C).

Effective diffusion coefficients for various solids encountered in food processing also are given by Schwartzberg and Chao (1982).

Concurrently, potato-slice-blanching research was scaled up to the large pilot scale. In the pilot plant (Fig. 7-2) work discussed previously, the water flow pattern through the blancher nearly approximated complete backmixing (CSTR), and was correlated by a model representing the flow as 94% backmixed and 6% plug flow. At the larger scale (875 liter) less thorough mixing in the blancher was expected, more nearly duplicating commercial scale blanchers.

To determine the flow pattern in the larger pilot plant unit operating at steady state, a lactose tracer solution was fed at time equal zero, and the exit concentration was monitored by refractive index as a function of time. Using the models given in Levenspiel (1972), the flow was found to consist of 78% backmixed and 24% plug flow, as shown in Figure 7-4 (Kozempel et al., 1985).

To calculate leaching using the plug flow model, the blancher was arbitrarily separated into *mixing zones* with the width of the zones equal to the pitch of the screw used to move the potatoes through the blancher. The temperature of a single site in each zone was monitored and was found to be *uniform* within the zones. The blanch model was applied sequentially to each zone at the temperature of the zone using the exit concentration of the previous zone as the inlet concentration. This plug flow model was applied to 24% of the flow stream in the blancher.

Temperature and concentration are uniform in backmixing. Therefore, the blancher exit temperature was used for the backmix contribution, and the backmix model was applied to 76% of the flow in the blancher (Kozempel et al., 1985).

Another identified complication is piece size. In the pilot plant tests piece size was fairly uniform. Figure 7-5 illustrates a typical piece-size distribution encountered in the pilot plant process. Since distribution is somewhat random and the increased difficulty of making iterative calculations for each piece size does not warrant the increased effort, an average piece size was used.

SOLUBILITY

Separation by fractional crystallization depends upon the relative solubility of the solutes. In simple cases, each solute is considered independent of the other solutes. A solubility curve must be established to design a fractional crystallization process. Over small temperature intervals, for solutions in which there

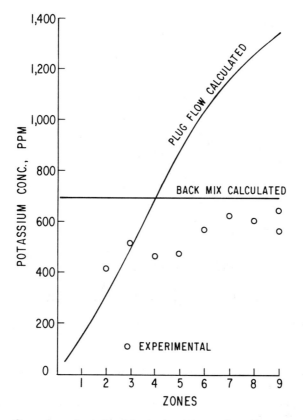

FIGURE 7-4.   Comparison of experimental potassium concentration with concentration calculated assuming back mix or plug flow in pilot plant.

is no change in the nature of the solute, the solubility can be correlated using equation (7-6) (Kozempel, 1971).

$$\ln m = -(\delta H/R)(1/T) + c \qquad (7\text{-}6)$$

In some food systems, for example, in fat-containing systems, the interaction of the various solutes or individual fractions may be very important. An example of this is seen in the process (Fig. 7-6) to fractionally crystallize beef tallow into a cocoa butter substitute. The solubility of beef tallow in acetone, curve A in Figure 7-7 (Kozempel et al., 1981b), was determined to develop a process for fractionating tallow to obtain a cocoa butter substitute. However, in attempting to establish the temperature for fractionating (crystallizing) the

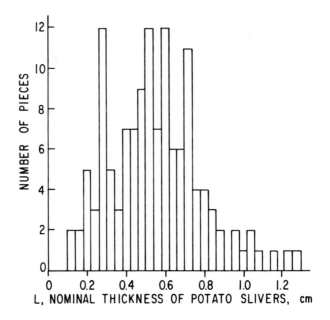

FIGURE 7-5.   Distribution of piece sizes in a commercial hot-water blancher.

stearine or high melting fat fraction, the process did not follow the solubility curve as expected.

For example, starting with a tallow concentration of 0.093, crystallization began at 29°C. The concentration of the filtrate did not follow solubility curve A, but followed line $B1$ instead (Fig. 7-7).

Starting with a different initial concentration, the filtrate concentration (crystallization) followed a different line, for example, $B2$. Although it is well known that solubility depends on the temperature of a system and not the initial concentration, these data gave every indication that, in this system, solubility depends on the initial concentration.

The explanation of this apparently anomalous solubility behavior stems from the nature of beef tallow. It is composed of many different triglycerides. Curve A establishes the solubility of the highest melting or least soluble component of the original mixture. As crystallization proceeds, the filtrate solution is stripped of the highest melting crystallizing fat, and curve A no longer applies to the liquid. Instead, crystallization is governed by a different solubility curve corresponding to a new least soluble fraction (curve A', Fig. 7-7). As this component crystallization continues, solubility becomes governed by the next least soluble component. In essence, an operating line is established by the intersections of the temperature with the family of solubility curves. It is reasonable to

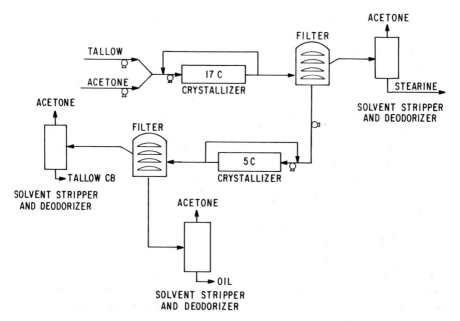

FIGURE 7-6. Process flow sheet for fractionating tallow.

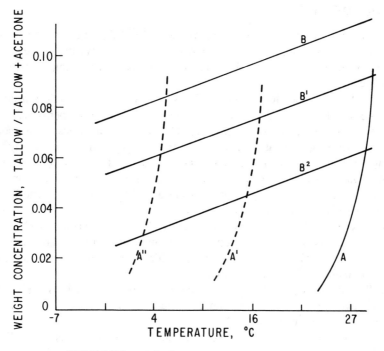

FIGURE 7-7. Solubility diagram for tallow in acetone.

expect that the various solubility curves and various operating lines would be parallel, since there is just a slight difference in the components.

The changing composition problem can be overcome by a systematic experimental procedure. First, an initial solubility curve must be determined. Next, one of the operating lines must be established by successively crystallizing a fat solution. From these curves, other operating lines are established parallel to the first, and the process is designed using the family of operating lines and the initial solubility curve.

In the cocoa butter substitute study, the solution was crystallized in acetone until the temperature reached 16–18°C, which separated the stearin fraction, a fraction useful as shortening or hard butter. This crystallization process nominally corresponded to a path defined by the intersection of the operating line with solubility curve A'.

Further cooling to a nominal 5°C produced a crystallized fraction with composition and physical properties almost identical to cocoa butter. This crystallization nominally corresponded to the intersection of the operating line with solubility curve A''. The filtrate was a liquid at room temperature, useful as a deep-fat frying oil.

DRUM DRYING

The mathematical treatment of *drum drying* was considered impractical before the advent of the computer. Attempts to formulate and solve the mathematical equations failed because of the great number and interrelationships of the variables.

Instant mashed potatoes or potato flakes are dried on a single drum dryer. The pretreatment of the potatoes has a strong influence on the performance of the dryer and the quality of the product. Therefore, it was necessary to model the preprocessing steps as well. Not only is every piece of potato slightly different (size, moisture content, starch composition, texture), but there are significant variations in cooked potato properties (per piece and within the mash). Also, the drum surface is not perfectly uniform. There are irregularities, hot spots, and end effects (at the sides of the drum where the side of the heated surface is exposed to the air and not coated with drying mashed potato).

Figure 7-8 is a sketch of the side view of a single drum dryer. Mash is loaded onto the drum surface at zero degrees of rotation and the dried flakes are doctored off at about 295 degrees of rotation. The pilot plant unit has four spreader rolls. The basic equation describing the drying can be developed from an energy balance over an infinitesimal volume, Figure 7-9, of mash of cross section L and r by δx (Kozempel et al., 1986).

The flow rate of mash is $dP/d\theta$ (kilogram mash/hour). Thus, $d^2P/d\theta^2$ is the change in mash flow rate, or the drying rate. The drying rate is expressed

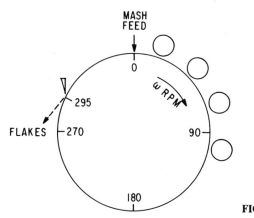

FIGURE 7-8.   Side view of drum dryer.

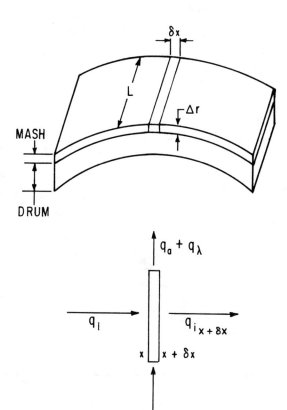

FIGURE 7-9.   Diagram of drum and mash sheet for modeling drum dryer.

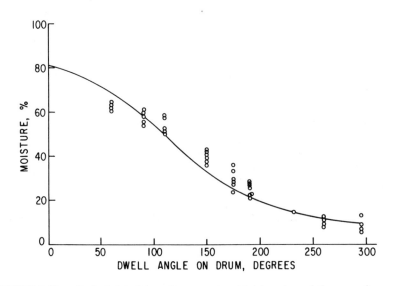

FIGURE 7-10. Typical plot of drum dryer experimental data and correlation curve for moisture vs. position on drum surface.

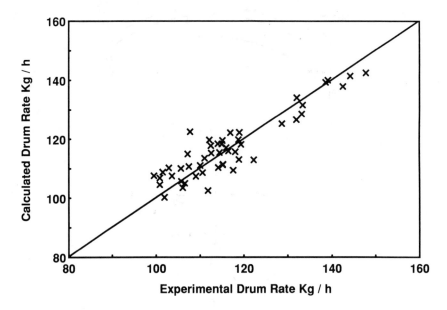

FIGURE 7-11. Plot of calculated vs. experimental drum dryer rate for two processing seasons.

as

$$\frac{-d^2P}{d\theta^2} = \frac{hML(T_s - T) - hM\left(\dfrac{T_sT_f}{T_f - T_a}\right)L(T - T_a)}{\dfrac{\lambda + CpT}{2\pi R\omega}} \qquad (7\text{-}7)$$

Solving this equation requires the solutions of several auxiliary equations. The temperature of the mash can be calculated from the wet-bulb temperature, the equilibrium mash temperature, and the moisture as in equation (7-8).

$$T = T_w + (T_f - T_w)\exp(-cM) \qquad (7\text{-}8)$$

The equilibrium mash temperature depends on the wet-bulb temperature and steam pressure

$$T_f = T_w + 99.11 + 5.12 \times 10^{-5}\ (\text{steam pressure}) \qquad (7\text{-}9)$$

In addition, to evaluate equation (7-7), the initial condition for the flow rate $(dP/d\theta)_0$, which is the capacity of the drum dryer, is needed. The mash flow rate depends on drum speed (rpm), steam pressure, number of spreader rolls, degree of cooking, starch content, bulk density of the dry potato, and the texture of the raw potato (Kozempel et al., 1990).

By using a *Hook–Jeeves pattern search* for the best values of the coefficients, a model that successfully predicts the moisture and capacity (drum dryer feed rate or $(dP/d\theta)_0$) was developed for drum drying potato flakes, as shown in Figures 7-10 and 7-11.

References

Balaban, M. 1988. Comparison of models for simultaneous heat and moisture transfer in food with and without volume change. Paper presented at the AIChE National Meeting, Denver, Colo., August 21–24.

Kozempel, M. F. 1971. Viscosity, solubility, density of isopropenyl stearate and isopropenyl acetate. *J. Chem. Eng. Data* **16**(3): 345–346.

Kozempel, M. F., J. F. Sullivan, and J. C. Craig, Jr. 1981a. Model for blanching potatoes and other vegetables. *Lebensmittel-Wissenshaft u.-Technol.* **14**:331–335.

Kozempel, M. F., J. C. Craig, Jr., W. K. Heiland, S. Elias, and N. C. Aceto. 1981b. Development of a continuous process to obtain a confectionery fat from tallow: Final status. *J. Am. Oil Chem. Soc.* **58**(10):921–925.

Kozempel, M. F., J. P. Sullivan, and J. C. Craig, Jr. 1985. Modeling and simulating commercial hot water blanching of potatoes. *Amer. Potato J.* **62**:69–82.

Kozempel, M. F., J. F. Sullivan, J. C. Craig, Jr., and W. K. Heiland. 1986. Drum drying potato flakes—A predictive model. *Lebensmittel-Wissenshaft u.-Technol.* **19**:193–197.

Kozempel, M. F., P. M. Tomasula, J. C. Craig, Jr., and M. J. Kurantz. 1990. Effect of potato composition on drum dryer capacity. *Lebensmittel-Wissenshaft u-Technol.* **123**:312–316.

Levenspiel, O. 1972. *Chemical Reaction Engineering*, 2d ed. New York: Wiley.

Newman, A. B. 1931. The drying of porous solids: Diffusion calculations. *Trans. Amer. Inst. Chem. Eng.* **27**:310–333.

Rotstein, E. 1990. Drying of foods. In *Biotechnology and Food Process Engineering*, H. G. Schwartzberg and M. A. Rao, eds. New York: Marcel Dekker.

Schwartzberg, H. G., and R. Y. Chao. 1982. Solute diffusivities in leaching processes. *Food Technol.:* 73–86.

Singh, R. P. 1982. Thermal diffusivity in food processing. *Food Technol.:* 87–91.

Talburt, W. F., and O. Smith. 1987. In *Potato Processing*, 4th ed. New York: Van Nostrand Reinhold.

Tomasula, P., and M. F. Kozempel. 1989. Diffusion coefficients of glucose, potassium, and magnesium in Maine Russet Burbank and Maine Katahdin potatoes from 45 to 90°C. *J. Food Sci.* **54**:985–989, 1046.

8

Rheology of Cheese

Michael H. Tunick and Edward J. Nolan

WHY RHEOLOGY IS USED IN CHEESE ANALYSIS

The instrumental measurement of the rheological properties of cheese is performed for two reasons: as a *quality control* method for cheesemakers, and as a technique for scientists to study *cheese structure*. The rheological properties of cheese can be as important as flavor, and are a large part of the total score awarded by the cheese grader (Farkye and Fox, 1990). Consequently, an objective instrumental method of determining rheological properties of cheese would be quite valuable. This also holds true for scientific studies of cheese. The pH at whey draining affects the size and shape of protein aggregates and has a large effect on the textural properties of cheese (Fox, Lucey, and Cogan, 1990). Low-pH curds result in a crumbly texture, such as in Cheshire cheese, whereas high-pH curds lead to more elastic cheeses, such as Swiss or Gouda. The extent of proteolysis and the amounts of water, protein, fat, and salt in cheese also affect texture, resulting in profound differences between cheese types. Research into the origins of cheese texture is an important part of dairy science, and investigations into cheese rheology have been conducted for over half a century. Correlating the results of various instrumental tests to each other and to subjective evaluations by sensory panels and consumers is a goal that is constantly being strived for (Szczesniak, 1987).

BRIEF HISTORY OF RHEOLOGICAL MEASUREMENTS OF CHEESE

The three general categories of food texture measurement are empirical, imitative, and fundamental tests (Scott Blair, 1958). The idea behind empirical tests is the measurement of parameters that experience indicates are related to texture. The first measurements of cheese rheology were empirical tests performed by hand without instruments: the cheese grader would press into the surface with his thumb to judge firmness and elasticity. Simple instruments followed, such as the ball compressor, in which a hemisphere presses into the surface of the specimen, and the depth of compression and amount of recovery with time is noted (Caffyn and Baron, 1947; Cox and Baron, 1955). Another instrument, the penetrometer, produces force measurements as a needle is pushed into the cheese (Prentice, 1987). A third device, the Cherry-Burrell Curd Tension Meter, measures the resistance to the passage of a wire through cheese curd or a block of cheese (Emmons and Price, 1959). Other tests include rate of penetration of a standard borer under a standard load, pressure on a standard cheese skewer pushed into cheese by hand, and rate of increase of indentation caused by increasing load on a sphere (Scott Blair and Baron, 1949). Szczesniak (1963) reviewed many of these devices. With all of these methods, several measurements must be taken over the sample surface in order to obtain an accurate average. The results can be useful if only one type of cheese is being tested, but these instruments are not always satisfactory for rigorous theoretical studies due to the arbitrary test conditions (Voisey, 1976).

In the 1930s, the first instrument to examine texture by imitating the chewing of food was developed: the Volodkevich bite tenderometer. A wedge, imitating teeth on the upper jaw, was brought down toward a second wedge, exerting a biting and squeezing action on the sample held between the wedges (Volodkevich, 1938). The next significant development in imitative tests came in the mid-1950s with the MIT denture tenderometer, which contained motorized dentures and strain gauges (Proctor et al., 1955). In the early 1960s, the General Foods Texturometer made its appearance (Friedman, Whitney, and Szczesniak, 1963). This device compresses a bite-sized sample to 25% of its original height, releases, and repeats, thus imitating jaw movement. Its strain gauges and a strip chart recorder produce a force–time curve from which a texture profile analysis (TPA) can be derived. Several years later, Bourne adapted the Instron Universal Testing Machine for TPA studies (Bourne, 1968), and the first studies of cheese with this instrument soon followed (Shama and Sherman, 1973a, 1973b). Theoretical analyses may be difficult when using imitative tests due to the various motions and stresses involved (Voisey, 1976).

Fundamental tests measure rheological properties such as viscous and elastic moduli, with the intention of relating the nature of the specimen to basic rheological models. Specimens used in these experiments are of a specific shape and are deformed in a specific manner; therefore, all of the test parameters are known, the results can be analyzed systematically, and predictions are more readily made. The earliest fundamental tests of cheese were performed in the 1930s when Davis compressed cylinders of four types of English cheese under constant load (Davis, 1937). These force-compression tests are the forerunners of much of the serious cheese rheology research in recent years. Scott Blair and his colleagues performed a great deal of fundamental work on cheese and other foods, emphasizing viscoelastic properties, which led to an increase in applications of rheological theory to food analysis (Bagley and Christianson, 1987).

METHODS USED

Fundamental Experiments

Force-Compression
Cheese presents classic viscoelastic behavior, that is, somewhere between the response of a viscous material and an elastic material. The theories behind fundamental measurements are covered by Collyer and Clegg (1988), with applications to food being described by Shoemaker, Lewis, and Tamura (1987) and Bagley and Christianson (1987). In fundamental rheological tests of cheese, there are several parameters that are measured, and several others that are calculated from them. The *stress* σ applied to a small sample, typically, a cylinder, is defined as the force per unit area of the specimen. Stress applied in a downward direction is known as *uniaxial compression*. The deformation resulting from uniaxial compression will cause a slight bulging of the specimen, and the original height of the sample h_0 will be changed by an amount Δh. The change $\Delta h/h_0$ is defined as *engineering strain*, a simplification of the more widely used *Hencky strain in compression*, defined by $-\ln(h/h_0)$ (Ward, 1971; Chatraei, Makosko, and Winter, 1981).

During force-compression tests, one usually lubricates the sample–platen interface to minimize frictional effects. The stress σ_L (the subscript L refers to the lubricated condition) at a strain $\epsilon = \Delta h/h_0$ is calculated on the assumption that a cylinder of original radius r_0 and height h_0 will be deformed to a cylinder of radius r and height h. Assuming no volume change occurs,

$$\pi r_0^2 h_0 = \pi r^2 h,$$

and the resulting stress is given by

$$\sigma_L = \frac{F}{\pi r^2} = \frac{F}{\pi r_0^2}\frac{h}{h_0}$$

where F is the total force applied to the sample. If the specimens are bonded to the platens, the sample cross-sectional area in contact with the platens is constant, and therefore the stress (when bonded) is given by

$$\sigma_b = \frac{F}{\pi r_0^2}$$

where the subscript b refers to the bonded state (Casiraghi, Bagley, and Christianson, 1985). Christianson, Casiraghi, and Bagley (1985) show that a modified form of an equation proposed by Gent and Lindley (1959) may then be used to compare bonded and lubricated results. This equation yields a corrected bonded stress

$$\sigma_{bc} = \sigma_b \left(1 + \frac{r_0^2}{2h^2}\right)^{-1}$$

Casiraghi, Bagley, and Christianson (1985) found that this relation is very satisfactory for Mozzarella cheese, up to about 60% deformation; the structure collapses above 60% in the bonded sample, invalidating the comparison. The aforementioned correction also gave good results in the case of Cheddar cheese under uniaxial compression before structural collapse, which occurred between strains of 0.4–0.8.

When the structure begins to break down, an inflection point appears in the *force–strain curve* (Fig. 8-1). Further along the curve, the structure collapses faster than the stress buildup. We define the stress at the point of fracture as the *yield point*.

Creep

Further experimentation that may be conducted to characterize cheese involves the application of a constant stress imposed suddenly at time zero and held constant thereafter; the resulting deformation strain is then measured as a function of time. The corresponding plot is called a *creep curve*. When the force is removed, strain decreases, producing a *recovery curve*. Recovery from this compression is known as *relaxation*. A typical *creep–relaxation curve* for Mozzarella is shown in Figure 8-2.

If the cheese behaves as a linear viscoelastic solid during the creep experiment, the creep strain ϵ_{xy} increases with time t, and is related to the stress σ by σ_{xy}

$$\epsilon_{xy} = J(t)\,\sigma_{xy}$$

where $J(t)$ is the *creep compliance function*. The subscript x refers to the direction of the perpendicular to the plane of the stress, and y refers to the direction of the stress component. It is possible to rearrange the creep compliance function into three parts, each of which corresponds to particular phases of creep: instantaneous deformation, retarded elastic deformation, and flow, that is,

$$J(t) = J_0 + J\Phi(t) + \frac{t}{\eta_3}$$

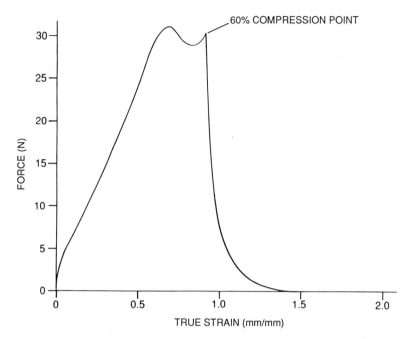

FIGURE 8-1   Force–strain curve of stirred curd Cheddar in uniaxial compression.

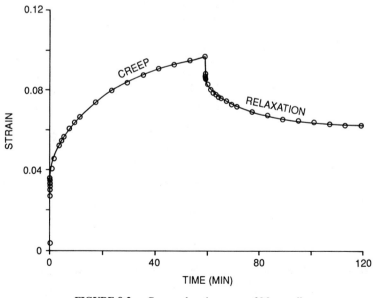

FIGURE 8-2   Creep–relaxation curve of Mozzarella.

where J_0 is compliance at time zero, J is the *decay compliance*, $\Phi(t)$ is a *retardation function* with initial value $\Phi(0) = 0$ and final value $\Phi(\infty) = 1$, and η_3 is the coefficient of viscosity in steady flow.

It is frequently adequate to represent experimental cheese creep data by describing the viscoelastic behavior using mechanical models (Fig. 8-3a and b). These models are comprised of elastic springs, which obey Hooke's law, and viscous dashpots, which obey the Newtonian viscosity law. Although the simplest possible models consist of a single spring and dashpot in series (Maxwell model) or in parallel (Voight or Kelvin model), neither is normally suitable for yielding a reasonable prediction of the behavior of cheese during creep. However, combining the springs and dashpots into a four-element model (Fig. 8-3c) gives a reasonable representation of cheese behavior during creep experiments (Chang et al., 1986). The total deformation represented by this combination of

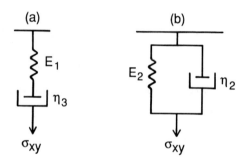

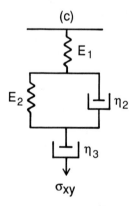

FIGURE 8-3   Mechanical models that describe viscoelastic behavior. (*a*) Maxwell model. (*b*) Voight or Kelvin model. (*c*) Four-element model.

Maxwell and Kelvin models is the sum of the deformation for each individual element:

$$\epsilon_{total} = \epsilon_1 + \epsilon_2 + \epsilon_3$$

where ϵ_1 is *instantaneous deformation*, ϵ_2 is *retarded elastic deformation*, and ϵ_3 is Newtonian flow (Lodge, 1964). Application of a stress σ_{xy} gives

$$\sigma_{xy} = E_1\epsilon_1 \qquad\qquad \text{(elastic)}$$

$$\sigma_{xy} = \frac{E_2\lambda d\epsilon_2}{dt} + E_2\epsilon_2 \quad (\lambda = \text{retardation time, } \eta_2/E_2)$$

$$\sigma_{xy} = \frac{\eta_3 d\epsilon_3}{dt} \qquad\qquad \text{(Newtonian flow)}$$

where E_1 and E_2 are constants. Differentiation of these equations with respect to time and subsequent elimination of the individual strains leads to the general equation for the four-element model during constant imposed stress σ_0:

$$\ddot{\epsilon} + \frac{\dot{\epsilon}}{\lambda} = \frac{\sigma_0}{\lambda\eta_3}$$

with the complementary solution

$$\epsilon = C_1 + C_2 \exp\frac{-t}{\lambda}$$

and particular integral

$$\epsilon = \frac{\sigma_0 t}{\eta_3}$$

The constants C_1 and C_2 are determined from the conditions

$$\epsilon = \frac{\sigma_0}{E_1} \qquad \text{at } t = 0$$

$$\epsilon \rightarrow \frac{\sigma_0}{E_2} \qquad \text{as } t \rightarrow \infty$$

finally giving the general solution

$$\epsilon = \frac{\sigma_0}{E_1} + \frac{\sigma_0[1 - \exp(-t/\lambda)]}{E_2} + \frac{\sigma_0 t}{\eta_3}$$

or in terms of compliance

$$\epsilon = J_0\sigma_0 + J_e\sigma_0\left[1 - \exp\left(\frac{-t}{\lambda}\right)\right] + \frac{\sigma_0 t}{\eta_3}$$

the subscript e representing the equilibrium value. Chang et al. (1986) analyzed the plotted creep data as follows: at zero time, the strain gives E_1; the extrapolated curve at long times gives a straight line that yields η_3; and the difference between the extrapolated portion of the creep curve and the creep curve itself gives rise to a curve from which E_2, η_2, and λ may be determined by curve fit.

Stress Relaxation

As noted earlier, relaxation ensues when a force is removed at the culmination of a creep experiment. A satisfactory method for modeling the behavior of cheese during this period is to employ a number of Maxwell elements in parallel, combined with a free spring. Ward (1971) has shown that the differential equation relating stress and strain in the deformation of viscoelastic materials is

$$a_0\sigma + \frac{a_1 d\sigma}{dt} + \frac{a_2 d^2\sigma}{dt^2} + \cdots = b_0 e + \frac{b_1 de}{dt} + \frac{b_2 d^2 e}{dt^2} + \cdots$$

When one includes only one or two terms of this equation, there results an equivalence to the corresponding behavior of elastic springs and viscous dashpots, obeying Hooke's law and Newton's law respectively. For example, stress relaxation occurs during a constant rate of strain; hence, de/dt is zero, as are the higher order derivatives of the strain. Consequently, an exponential response (Bland, 1960) is observed:

$$\sigma = \epsilon_0\left(E_0 + \sum_{m=1}^{n} E_m \exp\frac{-tE_m}{\eta_m}\right) u(t)$$

where η_m/E_m is the *relaxation time* and $u(t)$ is the *unit step function*:

$$u(t) = 0 \quad \text{when } t < 0$$

$$u(t) = 1 \quad \text{when } t > 0$$

The summation is based on the number of Maxwell elements in parallel, and E_0 is a free spring corresponding to the equilibrium value as $t \to \infty$. Generally, two or three terms are necessary to represent the experimental data obtained during relaxation. When using the foregoing model, the corresponding coefficients, the "elasticities" and "viscosities," are obtained by a curve fit to the experimental data, and may be used to compare particular cheese characteristics.

Steady Shear

In steady shear experiments, the sample is subjected to a *constant shear* in a rotational rheometer. The viscometer is used with a very small gap thickness to ensure that the velocity field is *linear*, or nearly so; this considerably simplifies the experimental determination. Transient stresses vanish during steady shear flow and the steady-state stresses depend only on the rate of shear $\dot{\gamma}$. A viscosity function dependent on shear rate, $\eta(\dot{\gamma})$, can be defined in analogy to the Newtonian viscosity (Bird, Armstrong, and Hassager, 1987):

$$\sigma_{yx} = -\eta(\dot{\gamma})\dot{\gamma}_{yx}$$

Normal stress coefficients $Z(\gamma)$ and $B(\gamma)$ are related by the equations

$$\sigma_{xx} - \sigma_{yy} = -Z(\dot{\gamma})(\dot{\gamma}_{yx})^2$$

$$\sigma_{yy} - \sigma_{zz} = -B(\dot{\gamma})(\dot{\gamma}_{yx})^2$$

Z, B, and $\eta(\dot{\gamma})$ completely determine the stress state in steady shear flow, and are consequently called *viscometric functions* (Vinogradov and Malkin, 1980). The quantities Z and B are called the *first and second normal stress coefficients*.

The viscosity functions are usually measured initially when examining oils or soft cheeses. Particular instrumentation must be used for measuring the first and second stresses in order to completely characterize the stress state. The fixture holding the sample may be two round parallel plates, *a cone and plate* (with the angle of the cone being less than about $3°$), or two concentric cylinders (a *Couette configuration*). One piece of the fixture—the lower plate, the cone, or a cylinder—rotates at a constant angular velocity while the other piece is stationary. The resistance of the specimen to the steady shear is measured by a transducer.

Parallel plates allow for samples of different thickness and for testing for sample slip. Cone and plate fixtures are useful for transient measurements such as stress relaxation and give a constant shear rate across the entire sample geometry. Dickie and Kokini (1982), for instance, measured shear stress and time-dependent flow of whipped cream cheese under steady shear with a cone and plate fixture. Concentric cylinders are best used for low-viscosity samples, such as cheese spreads.

Small Amplitude Oscillatory Shear

In *oscillatory shear* experiments, stress and strain vary *harmonically* with time. When stress and strain rates are small enough to stay within the linear range of viscoelasticity, the applicable mathematical equations are relatively simple. This type of technique is often described as *mechanical spectroscopy*, since it is analogous to other spectroscopic methods. Small-amplitude oscillatory shear measurements determine the transient response of a sample contained between two parallel plates separated by a distance h. The lower plate undergoes small-amplitude oscillations of frequency ω. The velocity profile is very nearly linear if

$$\frac{\omega \rho h^2}{2\eta_0} \ll 1$$

where ρ is the density and η_0 is the zero shear rate viscosity (Bird, Armstrong, and Hassager, 1987). During small-amplitude oscillatory shear measurements on cheese, the shear strain γ and shear rate $\dot{\gamma}$ are related by

$$\gamma = \mathring{\gamma} \sin(\omega t)$$

$$\dot{\gamma} = \mathring{\gamma} \omega \cos(\omega t)$$

where $\mathring{\gamma}$ is the amplitude.

The shear stress is *in phase* with the shear rate in Newtonian fluids, but not in cheese or most other foods. The stress of an ideally elastic material (Fig. 8-4) depends on the degree of deformation, which means that stress and strain are in phase. The stress of an ideally viscous material depends on deformation velocity; at maximum strain, the strain velocity is zero, since the oscillating system is changing direction. Therefore, stress and strain are 90° out of phase. This *phase angle*, δ, will lie between 0° and 90° in a viscoelastic substance. In the linear viscoelastic region the stress is related to the strain and a *frequency-dependent modulus M*:

$$\sigma = -M(\omega) \mathring{\gamma} \sin(\omega t + \delta)$$

where $0 \le \delta \le \pi/2$, and identically

$$\sigma = -M(\omega)\mathring{\gamma} [\sin(\omega t) \underline{\cos(\delta)} + \underline{\sin(\delta)} \cos(\omega t)].$$

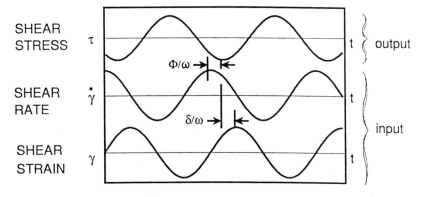

SHEAR
STRESS τ t   output

Φ/ω

SHEAR
RATE $\dot{\gamma}$ t   input

δ/ω

SHEAR
STRAIN γ t

FIGURE 8-4   Waveforms describing small-amplitude oscillatory shear (based on Lodge 1964).

The phase angle δ may be expressed in terms of complex moduli, as shown below:

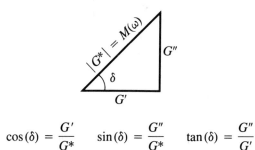

$$\cos(\delta) = \frac{G'}{G^*} \qquad \sin(\delta) = \frac{G''}{G^*} \qquad \tan(\delta) = \frac{G''}{G'}$$

leading to the equation

$$|G^*|^2 = (G')^2 + (G'')^2$$

The absolute value of the complex modulus G^*, which provides all of the information about the viscoelasticity of a sample, equals the maximum stress divided by maximum strain. The *elastic modulus* (or *storage* modulus) G' is $|G^*|\cos\delta$, and is a measure of the energy stored in the sample. The *viscous modulus* (or *loss* modulus) G'' is $|G^*|\sin\delta$, a measure of the energy lost as heat.

The viscosity coefficient may be determined in an analogous manner. The shear stress is related to the shear rate by

$$\sigma = -N(\omega)\dot{\gamma}$$

where $N(\omega)$ is a complex coefficient of viscosity. The *phase shift* Φ is related to the phase angle by

$$\Phi = \frac{\pi}{2} - \delta$$

In turn,

$$\sigma = -N(\omega)(\dot{\gamma})^\circ \cos(\omega t - \Phi)$$

where $0 \le \Phi \le \pi/2$. The shear rate amplitude is $(\dot{\gamma})^\circ$, so identically,

$$\sigma = -N(\omega)(\dot{\gamma})^\circ \, [\cos(\omega t) \, \underline{\cos(\Phi)} + \underline{\sin(\Phi)} \, \sin(\omega t)]$$

As can be seen below:

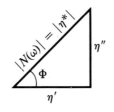

$$\cos(\Phi) = \frac{\eta'}{|\eta^*|} = \frac{\eta'}{|N(\omega)|} \equiv \sin(\delta)$$

$$\sin(\Phi) = \frac{\eta''}{|\eta^*|} = \frac{\eta''}{|N(\omega)|} \equiv \cos(\delta)$$

and

$$\tan \Phi = \frac{\eta''}{\eta'} \equiv \cot(\delta)$$

leading to the equation

$$|\eta^*|^2 = (\eta')^2 + (\eta'')^2$$

which gives the relationship of the complex viscosity η^* to the component viscosities η' and η''. These viscosities are related to the storage and loss moduli by

$$G' = \omega\eta'' \quad \text{and} \quad G'' = \omega\eta'$$

and are the relationships most often used by cheese rheologists.

The theories behind viscoelasticity are thoroughly covered by Ferry (1980), Christensen (1982), Aklonis and MacKnight (1983), and Sperling (1986). The measurement of viscoelasticity is performed under *dynamic* conditions, with the specimen undergoing *sinusoidally oscillating* deformation. Photographs of an oscillatory shear rheometer are shown in Figure 8-5. A study of Mozzarella on this apparatus by Nolan, Holsinger, and Shieh (1989) included tests for slippage caused by milk fat diffusing to the specimen surface. Differences in stress waveforms with two different sample thicknesses at the same strain and frequency provide evidence for slip, which was eliminated by bonding the specimen to the plates with cyanoacrylate glue.

The linear viscoelastic range of a material may be found by measuring the moduli under varying strain while holding the frequency constant—a *strain sweep* (Fig. 8-6). Above this range, the structure of the sample breaks down. Information about the moduli at constant strain may be obtained by varying the frequency—a *frequency sweep*. The strain selected must be in the linear viscoelastic region.

Texture Profile Analysis

The theory behind the methods used in TPA has been reviewed by Voisey (1976), Voisey and deMan (1976), and Prentice (1987). The General Foods Texturometer was the first instrument that could produce texture profiles of food, but the Instron Universal Testing Machine has been the instrument of choice by most rheologists since it was first used to analyze food in 1968 (Breene, 1975).

In the General Foods Texturometer, a sample is placed on a flat plate (or a shallow dish if the sample is a liquid) and a plunger is brought down onto it. The plunger decelerates as it reaches the end of the compression stroke, stops, and then accelerates upward. This produces a *sinusoidal deformation* motion that closely approximates the action of the human jaw, but also causes the viscoelastic reaction of the sample to vary during the analysis (Voisey and deMan, 1976). A force–time curve is produced, but the variable speed prevents conversion to force–distance (Bourne, 1976). In the Instron machine, a crosshead is sent down a vertical column, causing a flat plate to deform a specimen that has been placed on a lower plate. The Instron reaches the end of the compression stroke at full speed, stops abruptly, and accelerates upward at full speed, producing sharper peaks than the Texturometer. In addition, the constant speed leads to both force–time and force–distance curves; the work done can then be calculated since work is a force–distance integral (Bourne, 1976). Photographs of an Instron are shown in Figure 8-7.

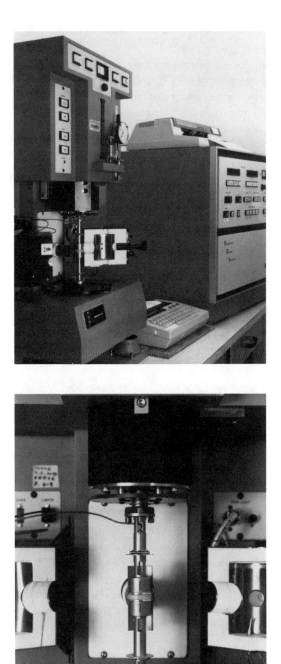

FIGURE 8-5 *Top:* Rheometrics RDA-700 oscillatory shear rheometer. *Bottom:* Cheese sample held between parallel plates.

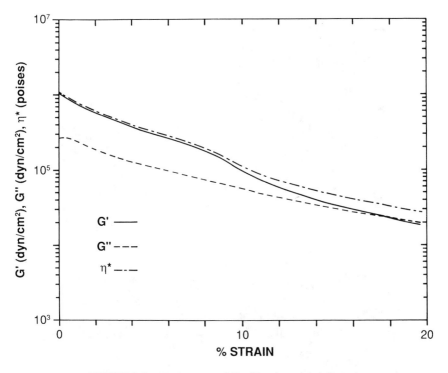

FIGURE 8-6   Strain sweep of Cheshire cheese aged 60 weeks.

Seven textural parameters have been extracted from the force–time curve obtained in TPA (Bourne, 1978):

1. *Fracturability* (originally brittleness), the force at the first significant break in the curve. The shoulder on the first peak in the curve in Figure 8-8 is the point to which fracturability is measured.
2. *Hardness*, the maximum force during the first compression cycle, or "first bite." This corresponds to the top of the first peak in Figure 8-8.
3. *Adhesiveness*, the force area of any negative peak following the first compression cycle (visible in Fig. 8-8). This variable represents the work required to pull the plunger or plate away from the sample.
4. *Cohesiveness*, the (dimensionless) ratio of the positive force area of the second compression to that of the first.
5. *Springiness* (originally elasticity), the height that the specimen recovers between the end of the first bite and the start of the second. In Figure 8-8, springiness is measured along the baseline from the start of the second peak

FIGURE 8-7   *Top:* Instron Universal Testing Machine. *Bottom:* Cheese sample held between plates.

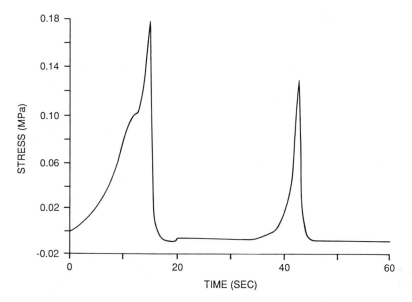

FIGURE 8-8   TPA curve of Mozzarella, obtained from an Instron instrument.

to the spot directly under the top of the peak. The measurement is then converted from seconds to units of length.

6. *Gumminess*, the product of hardness and cohesiveness.
7. *Chewiness*, the product of springiness and gumminess.

The Instron can also be used to measure stress relaxation, yield points, and other rheological parameters.

ANALYSES OF CHEESE VARIETIES

English Cheeses

Much rheological research has dealt with Cheddar and other English varieties of cheese, starting with Davis in the 1930s (Davis, 1937). He compressed cylinders of Cheddar, Cheshire, Lancashire, and Leicester under loads of 50, 100, and 200 g, and the resulting deformation-time curves were used to calculate moduli, viscosities, and relaxation times. Ball compressor tests on Cheshire and Cheddar were compared to graders' scores in a four-year study by Wearmouth (1954). More recently, force-compression tests using an Instron Universal Testing Machine have been run on Cheddar by several groups. Information obtained includes hardness, chewiness, springiness, and adhesiveness (Lee, Imoto, and

Rha, 1978); stress-strain relationships (Rosenau, Calzada, and Peleg, 1978; Casiraghi, Bagley, and Christianson, 1985); compression ratios (Imoto, Lee, and Rha, 1979); yield points (Dickinson and Goulding, 1980); and firmness and elasticity relationships with fat content (Emmons et al., 1980).

Creamer and Olson (1982) used an MTS Tensile Testing Machine to equate force-compression and yield point with casein proteolysis. They found that yield strain decreased linearly with the logarithm of days aged. Fedrick and Dulley (1984) used an Instron to determine that hardness and springiness decreased, and fracturability increased, with proteolysis. Amantea, Skura, and Nakai (1986) also compared ripening Cheddar samples with an Instron, finding that force-compression relationships are most affected by proteolysis, age, and type of starter culture.

An Instron compression test was used by Weaver, Kroger, and Thompson (1978) to determine toughness, hysteresis, stiffness, and degree of elasticity of Cheddar during ripening. Chen et al. (1979) attached a plunger to an Instron to penetrate samples (a "punch test"), and examined textural parameters of Cheddar and nine other varieties. Green et al. (1986) determined stiffness and work of fracture by cutting with a wire guillotine attached to an Instron. Nolan (1987) subjected stirred curd Cheddar to constant strain in an Instron and allowed the specimens to "relax," determining that a primary relaxation period occurred for 2 min, followed by a secondary period.

Fukushima, Taneya, and Sone (1964) obtained hardness data by slicing with a wire, yield stress by cone penetration, and other viscoelastic parameters by compression between parallel plates. Purkayastha et al. (1985) obtained a model of creep behavior of Cheddar using a homemade device.

Force-compression tests were used by Shama and Sherman (1973a, 1973b) to analyze Caerphilly, Double Gloucester, Lancashire, and White Stilton; by Carter and Sherman (1978) to analyze Leicester for firmness; and by Dickinson and Goulding (1980) to analyze yield points of Cheshire and Leicester. Green, Marshall, and Brooker (1985) compared ball compressor results for Cheddar and Cheshire to sensory panels.

Tunick et al. (1990) compared the viscoelastic properties of Cheddar and Cheshire with a *Rheometrics Dynamic Analyzer*. A straight line that followed the Arrhenius equation was obtained by plotting complex viscosity versus reciprocal of temperature at constant strain and frequency. The slope of the line was proportional to the activation energy, which was substantially different for the two varieties of cheese.

Italian Cheeses

The Instron has been used to examine Mozzarella hardness, chewiness, springiness, and adhesiveness (Lee, Imoto, and Rha, 1978); compression ratios (Imoto, Lee, and Rha, 1979); and stress-strain relationships (Casiraghi, Bagley, and Christianson, 1985). Smith, Rosenau, and Peleg (1980) measured flowability of melted Mozzarella with a piston-driven capillary rheometer. Taranto and Yang (1981) and Yang and Taranto (1982) obtained texture profiles from imitation Mozzarella cheese made from soybean oil. Cervantes, Lund, and Olson (1983) used an MTS Tensile Testing Machine for compression and deformation tests on Mozzarella. Nolan, Holsinger, and Shieh (1989) determined shear moduli and Arrhenius relationships of natural and imitation Mozzarella using a Rheometrics Dynamic Analyzer. Izutsu et al. (1990) determined apparent viscosity, relaxation time, and other parameters for Mozzarella as they relate to stretching of the curd.

Tunick et al. (1991) ran TPA on Mozzarella, varying fat content, moisture content, and type of storage. Hardness, gumminess, and chewiness all increased, and springiness decreased, when the fat or moisture contents were lowered. The same textural parameters were higher in one-week-old samples than in six-week-old or frozen samples, due to proteolysis.

Masi (1989) ran compression–relaxation tests on Galbanino cheese, and determined relaxation rate constants on seven other Italian cheeses. Masi and Addeo (1984) ran compression, stress relaxation, and TPA on Caciocavallo, Caciotta, Mozzarella (buffalo and cow milk), Provoloncino, Salamino, and Silano using a General Foods Texturometer. Casiraghi, Lucisano, and Pompei (1989) studied texture profiles of Grano Padino, Italico, Montasio, Pecorino, and Sbrinz.

Dutch Cheeses

Force-compression tests of Edam and Gouda were performed with an Instron by Shama and Sherman (1973a), who also investigated the stress relaxation characteristics of Edam (1973b). DeJong (1976) found that the firmness of Meshanger, as measured by the Instron, was related to breakdown of α_{s1}-casein to the α_{s1}-I peptide. Firmness of Gouda was evaluated by Instron force-compression tests by Culioli and Sherman (1976), who noted that samples near the edge of a block may require twice the stress for failure as samples at the center due to moisture gradients. Stress relaxation of Gouda was examined by Goh and Sherman (1987). Luyten (1988) used an Overload Dynamics material testing instrument to determine viscoelastic and fracture properties of Gouda as they varied with composition and aging. Fukushima, Taneya, and Sone (1964) analyzed rheological properties of Gouda in the same manner as they did with Cheddar.

French, German, and Swiss Cheeses

Compression and relaxation studies of Camembert were performed with an Instron by Mpagana and Hardy (1985, 1986). Yield point, deformation, and relaxation characteristics of Camembert and St. Paulin were examined with an Instron by Kfoury, Mpagana, and Hardy (1989). Several commercial acid-type French cheeses were analyzed for apparent viscosity and strain with a coaxial cylinder viscometer (Korolczuk and Mahaut, 1989), and for stress relaxation with a cone penetrometer (Korolczuk and Mahaut, 1990). Molander, Kristiansen, and Werner (1990) ran compression tests on Brie.

Eberhard and Flückiger (1981) differentiated between Appenzell, Emmental, Gruyère, Sbrinz, and Tilsit by examining their force-deformation curves with an Instron. Eberhard (1985) determined compression strength, deformation, and elasticity of Emmental as it aged. Lee, Imoto, and Rha (1978) and Imoto, Lee, and Rha (1979) ran force-deformation tests on Muenster and Swiss.

Cottage and Cream Cheeses

The textural evaluation of soft cheeses has been primarily by empirical means. An Instron was used to measure textural parameters by Lee, Imoto, and Rha (1978) and Imoto, Lee, and Rha (1979). An Instron was adapted to measure firmness of cottage cheese by Perry and Carroad (1980). Hori (1982) used a TOM compression testing machine to determine yield point data, stiffness, and deformability of cream cheese after freezing and thawing the curds. Dickie and Kokini (1982) measured shear stress under steady shear of whipped cream cheese.

Other Cheeses

Chen and Rosenberg (1977) devised a model simulating the yield behavior of American cheese using Instron data. Textural parameters of American cheese samples were examined by Lee, Imoto, and Rha (1978) and Imoto, Lee, and Rha (1979) using an Instron. Campanella et al. (1987) obtained an indication of American cheese meltability by compressing melting samples in an Instron and measuring elongational viscosity. Chang et al. (1986) developed a college laboratory experiment in which the viscoelasticity of processed cheese, especially in creep, was measured. Marshall (1990) measured work of fracture of processed cheese analogues by using a wire guillotine in an Instron in the same manner as Green et al. (1986). Fukushima, Sone, and Fukada (1965) determined viscosity and dynamic elastic moduli of processed cheese using a viscometer.

THE FUTURE

The factors affecting the rheological properties of cheese must be known if acceptable new products are to be developed. Many of the rheological tests that are performed in the cheese industry are empirical in nature, and more reliable data are needed along with standardized testing methods. Difficulties in obtaining representative samples from crumbly cheeses, such as Cheshire, or cheeses with eyes, such as Swiss, will have to be overcome. At present, there is no method capable of analyzing all of the parameters responsible for texture and mouthfeel of food. When such a technique is developed, rheological analyses will accurately predict the quality of cheese while providing insight into the chemistry of this popular food.

References

Aklonis, J. J., and W. J. MacKnight. 1983. *Introduction to Polymer Viscoelasticity*, 2d ed. New York: Wiley.

Amantea, G. F., B. J. Skura, and S. Nakai. 1986. Culture effects on ripening characteristics and rheological behavior of Cheddar cheese. *J. Food Sci.* **51**:912–918.

Bagley, E. B., and D. D. Christianson. 1987. Measurement and interpretation of rheological properties of foods. *Food Technol.* **41**(3):96–99.

Bird, R. D., R. C. Armstrong, and O. Hassager. 1987. *Dynamics of Polymeric Liquids*, vol. 1: *Fluid Mechanics*. New York: Wiley.

Bland, D. R. 1960. *The Theory of Linear Viscoelasticity*. New York: Pergamon.

Bourne, M. C. 1968. Texture profile of ripening pears. *J. Food Sci.* **33**:223–226.

Bourne, M. C. 1976. Interpretation of force curves from instrumental texture measurements. In *Rheology and Texture in Food Quality*, J. M. deMan, P. W. Voisey, V. F. Rasper, and D. W. Stanley, eds., 244–274. Westport, Conn.: AVI.

Bourne, M. C. 1978. Texture profile analysis. *Food Technol.* **32**:62–66, 72.

Breene, W. M. 1975. Application of texture profile analysis to instrumental food texture evaluation. *J. Texture Stud.* **6**:53–82.

Caffyn, J. E., and M. Baron. 1947. Scientific control in cheese making. *Dairyman* **64**:345, 347, 349.

Campanella, O. H., L. M. Popplewell, and J. R. Rosenau, and M. Peleg. 1987. Elongational viscosity measurements of melting American process cheese. *J. Food Sci.* **52**:1249–1251.

Carter, E. J. V., and P. Sherman. 1978. Evaluation of the firmness of Leicester cheese by compression tests with the Instron Universal Testing Machine. *J. Texture Stud.* **9**:311–324.

Casiraghi, E. M., E. B. Bagley, and D. D. Christianson. 1985. Behavior of Mozzarella, Cheddar, and processed cheese spread in lubricated and bonded uniaxial compression. *J. Texture Stud.* **16**:281–301.

Casiraghi, M., M. Lucisano, and C. Pompei. 1989. Correlation among instrumental texture, sensory texture and chemical composition of five Italian cheeses. *Ital. J. Food Sci.* **1**:53–63.

Cervantes, M. A., D. B. Lund, and N. F. Olson. 1983. Effects of salt concentration and freezing on Mozzarella cheese. *J. Dairy Sci.* **66**:204–213.

Chang, Y. S., J. S. Guo, Y. P. Lee, and L. H. Sperling. 1986. Viscoelasticity of cheese. *J. Chem. Educ.* **63**:1077–1078.

Chatraei, S., C. W. Makosko, and H. H. Winter, 1981. Lubricated squeezing flow: A new biaxial extensional rheometer. *J. Rheol.* **25**:433–443.

Chen, A. H., J. W. Larkin, C. J. Clark, and W. E. Irwin. 1979. Textural analysis of cheese. *J. Dairy Sci.* **62**:901–907.

Chen, Y., and J. Rosenberg. 1977. Nonlinear viscoelastic model containing a yield element for modeling a food material. *J. Texture Stud.* **8**:477–485.

Christensen, R. W. 1982. *Theory of Viscoelasticity. An Introduction*, 2d ed. New York: Academic Press.

Christianson, D. D., E. M. Casiraghi, and E. B. Bagley. 1985. Uniaxial compression of bonded and lubricated gels. *J. Rheol.* **29**:671–684.

Collyer, A. A., and D. W. Clegg. 1988. *Rheological Measurement.* New York: Elsevier.

Cox, C. P., and M. Baron. 1955. A variability study of firmness in cheese using the ball-compressor test. *J. Food Res.* **22**:386–390.

Creamer, L. K., and N. F. Olson. 1982. Rheological evaluation of maturing Cheddar cheese. *J. Food Sci.* **47**:631–636, 646.

Culioli, J., and P. Sherman. 1976. Evaluation of Gouda cheese firmness by compression tests. *J. Texture Stud.* **7**:353–372.

Davis, J. G. 1937. The rheology of cheese, butter and other milk products. *J. Dairy Res.* **8**:245–264.

DeJong, L. 1976. Protein breakdown in soft cheese and its relation to consistency. 1. Proteolysis and consistency of 'Noordhollandse Meshanger' cheese. *Neth. Milk Dairy J.* **30**:242–253.

Dickie, A. M., and J. L. Kokini. 1982. Use of the Bird-Leider equation in food rheology. *J. Food Proc. Eng.* **5**:157–182.

Dickinson, E., and I. C. Goulding. 1980. Yield behaviour of crumbly English cheeses in compression. *J. Texture Stud.* **11**:51–63.

Eberhard, P., and E. Flückiger. 1981. Rheological measurements on different kinds of cheese. *Schweitzer Milchztg.* **107**:23.

Eberhard, P. 1985. Rheological properties of some cheese varieties. I. Emmental cheese. *Schweitzer Milchztg. Forsch.* **14**:3–9.

Emmons, D. B., and W. V. Price. 1959. A curd firmness test for cottage cheese. *J. Dairy Sci.* **42**:553–556.

Emmons, D. B., M. Kalab, E. Larmond, and R. J. Lowrie. 1980. Milk gel structure. X. Texture and microstructure in Cheddar cheese made from whole milk and from homogenized low-fat milk. *J. Texture Stud.* **11**:15–34.

Farkye, N. Y., and P. F. Fox. 1990. Objective indices of cheese ripening. *Trends Food Sci. Technol.* **1**:37–40.

Fedrick, I. A., and J. R. Dulley. 1984. Effect of elevated storage temperatures on the rheology of Cheddar cheese. *N. Z. J. Dairy Sci. Technol.* **19**:141–150.

Ferry, J. D. 1980. *Viscoelastic Properties of Polymers*, 3d ed. New York: Wiley.

Fox, P. F., J. A. Lucey, and T. M. Cogan. 1990. Glycolysis and related reactions during cheese manufacture and ripening. *Crit. Rev. Food Sci. Nutr.* **29**:237–253.

Friedman, H. H., J. E. Whitney, and A. S. Szczesniak. 1963. The Texturometer—A new instrument for objective texture measurement. *J. Food Sci.* **28**:390–396.

Fukushima, M., S. Taneya, and T. Sone. 1964. Viscoelasticity of cheese. *J. Soc. Mater. Sci., Japan* **13**:331–335.

Fukushima, M., T. Sone, and E. Fukada. 1965. Effect of moisture content on the viscoelasticity of cheese. *J. Soc. Mater. Sci.*, *Japan* **14**:270–273.

Gent, A. M., and P. B. Lindley. 1959. The compression of bonded and rubber blocks. *Proc. Inst. Mech. Eng.* **173**:111–122.

Goh, H. C., and P. Sherman. 1987. Influence of surface friction on the stress relaxation of Gouda cheese. *J. Texture Stud.* **18**:389–404.

Green, M. L., R. J. Marshall, and B. E. Brooker. 1985. Instrumental and textural assessment and fracture mechanisms of Cheddar and Cheshire cheeses. *J. Texture Stud.* **16**:351–361.

Green, M. L., K. R. Langley, R. J. Marshall, B. E. Brooker, A. Willis, and J. F. V. Vincent. 1986. Mechanical properties of cheese, cheese analogues and protein gels in relation to composition and microstructure. *Food Microstruct.* **5**:169–180.

Hori, T. 1982. Effects of freezing and thawing green curds before processing on the rheological properties of cream cheese. *J. Food Sci.* **47**:1811–1817.

Imoto, E. M., C.-H. Lee, and C. Rha. 1979. Effect of compression ratio on the mechanical properties of cheese. *J. Food Sci.* **44**:343–345.

Izutsu, T., M. Azuma, T. Kimura, and M. Fukushima. 1990. Rheological properties of Mozzarella cheese curds. *Nippon Kagaku Kaishi* **1990**:621–627.

Kfoury, M., M. Mpagana, and J. Hardy. 1989. Effect of cheese ripening on rheological properties of Camembert and Saint-Paulin cheeses. *Lait* **69**:137–149.

Korolczuk, J., and M. Mahaut. 1989. Viscosimetric studies in acid type cheese texture. *J. Texture Stud.* **20**:169–178.

Korolczuk, J., and M. Mahaut. 1990. Relaxation studies of acid type cheese texture by a constant speed cone penetrometric method. *J. Texture Stud.* **21**:107–122.

Lee, C.-H., E. M. Imoto, and C. Rha. 1978. Evaluation of cheese texture. *J. Food Sci.* **43**:1600–1605.

Lodge, A. S. 1964. *Elastic Liquids*, 113. New York: Academic Press.

Luyten, H. 1988. *The Rheological and Fracture Properties of Gouda Cheese.* Ph.D. Thesis, Wageningen Agricultural Univ., The Netherlands.

Marshall, R. J. 1990. Composition, structure, rheological properties, and sensory texture of processed cheese analogues. *J. Sci. Food Agric.* **50**:237–252.

Masi, P. 1989. Characteristics of history-dependent stress-relaxation behavior of cheeses. *J. Texture Stud.* **19**:373–388.

Masi, P., and F. Addeo. 1984. Rheological characteristics of some typical pasta filata cheeses. *Latte* **9**:658–671.

Molander, E., K. R. Kristiansen, and H. Werner. 1990. Instrumental and sensoric measurement of Brie texture. *Milchwissenschaft* **45**:589–593.

Mpagana, M., and J. Hardy. 1985. Compression and relaxation properties of soft cheeses. Effect of ripening. *Sci. Aliment.* **5**:91–96.

Mpagana, M., and J. Hardy. 1986. Effect of salting on some rheological properties of fresh Camembert cheese as measured by uniaxial compression. *Milchwissenschaft* **41**:210–213.

Nolan, E. J. 1987. Stress relaxation of stored stirred Cheddar curd. *J. Texture Stud.* **18**:273–280.

Nolan, E. J., V. H. Holsinger, and J. J. Shieh. 1989. Dynamic rheological properties of natural and imitation Mozzarella cheese. *J. Texture Stud.* **20**:179–189.

Perry, C. A., and P. A. Carroad. 1980. Instrument for texture of small curd cottage cheese and comparison to sensory evaluation. *J. Food Sci.* **45**:798–801.

Prentice, J. H. 1987. Cheese rheology. In *Cheese: Chemistry, Physics, and Microbiology*, P. F. Fox, ed., 299–314. New York: Elsevier.

Proctor, B. E., S. Davison, G. J. Malecki, and M. Welch. 1955. A recording strain-gage denture tenderometer for foods. I. Instrumental evaluation and initial tests. *Food Technol.* **9**:471–477.

Purkayastha, S., M. Peleg, E. A. Johnson, and M. D. Normand. 1985. A computer aided characterization of the compressive creep behavior of potato and Cheddar cheese. *J. Food Sci.* **50**:45–50, 55.

Rosenau, J. R., J. F. Calzada, and M. Peleg. 1978. Some rheological properties of a cheese-like product prepared by direct acidification. *J. Food Sci.* **43**:948–950, 953.

Scott Blair, G. W. 1958. Rheology in food research. *Adv. Food Res.* **8**:1–61.

Scott Blair, G. W., and M. Baron. 1949. The rheology of cheese-making. *Proc. 12th Int. Dairy Congr.* **2**:49–58.

Shama, F., and P. Sherman. 1973a. Evaluation of some textural properties of foods with the Instron Universal Testing Machine. *J. Texture Stud.* **4**:344–352.

Shama, F., and P. Sherman. 1973b. Stress relaxation during force-compression studies on foods with the Instron Universal Testing Machine and its implications. *J. Texture Stud.* **4**:353–362.

Shoemaker, C. F., J. I. Lewis, and M. S. Tamura. 1987. Instruments for rheological measurements of food. *Food Technol.* **41**(3):80–84.

Smith, C. E., J. R. Rosenau, and M. Peleg. 1980. Evaluation of the flowability of melted Mozzarella cheese by capillary rheometry. *J. Food Sci.* **45**:1142–1145.

Sperling, L. H. 1986. *Introduction to Physical Polymer Science*. New York: Wiley.

Szczesniak, A. S. 1963. Objective measurements of food texture. *J. Food Sci.* **28**:410–420.

Szczesniak, A. S. 1987. Correlating sensory with instrumental texture measurements—An overview of recent developments. *J. Texture Stud.* **18**:1–15.

Taranto, M. V., and C. S. T. Yang. 1981. Morphological and textural characterization of soybean Mozzarella cheese analogs. *Scan. Electron Microscop.* **3**:483–492.

Tunick, M. H., E. J. Nolan, J. J. Shieh, J. J. Basch, M. P. Thompson, B. E. Maleeff, and V. H. Holsinger. 1990. Cheddar and Cheshire cheese rheology. *J. Dairy Sci.* **73**:1671–1675.

Tunick, M. H., K. L. Mackey, P. W. Smith, and V. H. Holsinger. 1991. Effects of composition and storage on the texture of Mozzarella cheese. *Neth. Milk Dairy J.* **45**:117–125.

Vinogradov, G. V., and A. Ya. Malin. 1980. *Rheology of Polymers*. Moscow: Mir.

Voisey, P. W. 1976. Instrumental measurements of food texture. In *Rheology and Texture in Food Quality*, J. M. deMan, P. W. Voisey, V. F. Rasper, and D. W. Stanley, eds., 79–141. Westport, Conn.: AVI.

Voisey, P. W., and J. M. deMan. 1976. Applications of instruments for measuring food texture. In *Rheology and Texture in Food Quality*, J. M. deMan, P. W. Voisey, V. F. Rasper, and D. W. Stanley, eds., 142–243. Westport, Conn.: AVI.

Volodkevich, N. N. 1938. Apparatus for measurements of chewing resistance or tenderness of foodstuffs. *Food Res.* **3**:221–225.

Ward, I. M. 1971. *Mechanical Properties of Solid Polymers*, 2d ed. New York: Wiley.

Wearmouth, W. G. 1954. A further study of the relationship between the mechanical properties of Cheddar cheese and graders' judgments. *Dairy Ind.* **19**:213–217.

Weaver, J. C., M. Kroger, and M. P. Thompson. 1978. Free amino acid and rheological measurements on hydrolyzed lactose Cheddar cheese during ripening. *J. Food Sci.* **43**:579–583.

Yang, C. S. T., and M. V. Taranto. 1982. Textural properties of Mozzarella cheese analogues manufactured from soybeans. *J. Food Sci.* **47**:906–910.

9

Basic Aspects of Food Extrusion

Food extrusion is one of the established technologies in food processing. Basic understanding of food extrusion, which involves both physical and chemical processes is, however, much less advanced than either the applications or the computation/prediction of the extrusion process for *plastic* polymeric materials. The reasons for the latter limitations lie in both the diversity and complexity of the food materials that are extruded. The modeling of the food extrusion process does require a knowledge of the *flow behavior* (see also the section on flow in Chap. 1), or the rheology of the heated food doughs (or "melts"). As indicated in Chapter 1, most foods have non-Newtonian flow characteristics that complicate the numerical analysis of the extrusion process. Furthermore, the extrusion cooking of foods, in most cases, involves *nonisothermal* conditions, complex heat transfer processes through the dough, and *irreversible* (thermodynamic) processes (see also Chap. 2 for an explanation of basic thermodynamic concepts, terms, and chemical reactions).

As a first step, or rough approximation, in the numerical analysis/modeling of the food extrusion process one can consider the extrusion of Newtonian fluids under *isothermal* conditions, and neglect chemical reactions or other irreversible changes induced by extrusion cooking of the food dough. This approach (Harper, 1981) yields exact solutions of the flow equations for the extruder, and it is useful only as a general guide to approaching the more complex, real problems of food extrusion.

Recently, a second approach to the food extrusion problem was developed on an experimental basis. The experimental setup is such that the experimental extruder has a rheologically simple profile (such as a capillary rheometer), and the extrusion conditions are closely controlled to make them near-isothermal). Material changes caused by extrusion cooking are estimated by determining the relationship between the mechanical energy input and the degree of degradation of the material. The results of the extruder-fed rheometry and isothermal capillary rheometry are then compared. To the latter approach one could add the sophistication of dynamic mechanical thermal analysis (DMTA) in order to identify thermal transitions in the food doughs and their correlation with heat-induced changes in their flow properties. Furthermore, the complexity of the extrusion process requires, in addition to rheology, a detailed study by a combination of powerful physical and chemical techniques that were presented briefly in Chapter 5 (X-ray diffraction, NMR, ESR, and DSC/DTA). Various types of flow behavior in foods were discussed briefly in Chapter 1 and more extensively in Chapter 8. It will be instructive to consider first the extrusion process for Newtonian fluids under isothermal conditions, and then to proceed to more realistic models of foods such as pseudoplastic doughs and correction factors for "nonidealities" of the extrusion process.

An extruder can be analyzed or divided into a number of sections based on the processing step or the type of process occurring in that section. The food material is fed into the extruder through a feed hopper, for example, and the corresponding part of the extruder is called the *feed section*. A tapered, rotating screw (Fig. 9-1) moves the hydrated food dough and compresses it, causing it to heat up at the same time. The part of the extruder where the food dough is compressed and heated is called the *compression* (or *transition*) *section*. The gradual decrease in the distance between threads (flight depth), or the pitch, achieves compression in the transition section. This section occupies about half the length of the extruder. Next is the *metering section*, which is the part nearest to the discharge of the extruder and often has only very shallow intervals (Fig. 9-1). This reduced pitch causes the effective shear rate in the channel to increase to the maximum possible for a given rotation rate, N, of the extruder screw. Because of the higher shear rate in the metering section, the internal mixing is quite vigorous, and the conversion of mechanical work into heat, as a result of the substantial energy dissipation effects by viscous flow at the high shear rates, causes a sharp increase in the dough temperature. The temperature profile shows

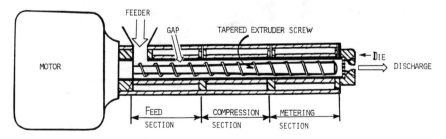

FIGURE 9-1.   Schematic diagram of a single-screw extruder showing the extruder sections.

an almost linear increase with time during the preconditioning/mixing of the dough, as well as a sharp peak to about 180°C just before the discharge at the die. The discharge is followed by a rapid decrease in temperature, where often expansion ("puffing") of the food dough occurs.

There are several advantages to a food extruder, like the one shown in Figure 9-2, for food processing: versatility in use; high productivity (continuous processing is possible with many high throughput industrial extruders); relatively low running costs; low overhead; high food product quality (as a result of the *high-temperature, short-time* [HTST] process involved that minimizes degradation of most food ingredients); and waste control (no pollutants). Some of the many applications of the food cooking extruder are for pasta (continuous process), ready-to-eat (RTE) cereal products, expanded corn collets or curls, texturized vegetable proteins confections, puffed corn meal, soup/gravy bases, and cookies or cracker-type shapes. There are several competing extruder designs in the marketplace; for example, both single-screw and twin-screw extruders are widely employed in commercial food production. The latter extruder

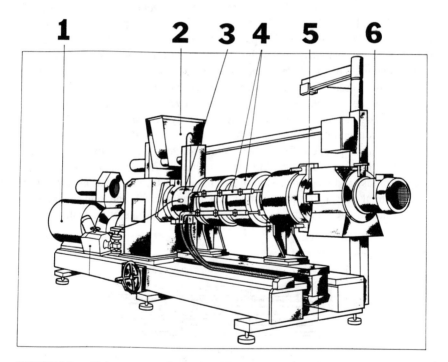

FIGURE 9-2. Twin-screw extruder employed for extrusion cooking of foods: (1) motor, (2) feeder, (3) barrel, (4) thermostatting system, (5) metering section, (6) die discharge zone.

is much harder to model because of the complex nature of the flow (vortexing) between the twin screws, but the twin screw extruder is preferred in diversified food companies because its design allows for more accurate control of the effective shear rate in the dough, as well as control of the residence time of the dough in the extruder barrel.

EXTRUSION OF NEWTONIAN MATERIALS UNDER ISOTHERMAL CONDITIONS

When looking at a single-screw food extruder in cross-section (Fig. 9-3), one sees a flighted Archimedes screw fitting tightly in a cylindrical barrel within which the screw can rotate. The flights push and compress the food material or dough, as well as mix and work the dough into a continuous paste. See Harper (1981) for a detailed presentation of the terminology of the various parts of an extruder.

To simplify the calculations of the flow in an extruder, one begins by considering laminar flow in a special type of extruder *capillary* that has a very narrow barrel, and the screw replaced by a simple *piston* (e.g., without flights; Fig. 9-4). Such a measuring instrument, which mimics the extruder in an oversimplified way, is called a *capillary rheometer*. With this geometry, it is often possible to obtain laminar, steady flow so that there is *no slippage* of the fluid at the capillary walls and the axial velocity of the fluid depends only on the distance from the axis of the capillary. Isothermal or nearly isothermal conditions are achieved relatively easily by thermostating the barrel. The use of relatively long tubes reduces significantly the *end effects* (Bagley, 1954) present in extruders.

VELOCITY PROFILE CALCULATION

The *velocity profile* of the fluid in such an extruder is readily calculated by considering the difference in pressure ΔP on the ends of a tiny cylinder of fluid inside the capillary. This difference in pressure is balanced by the applied shear stress, so that for a cylinder element of radius r, one has

$$\Delta P \cdot (\pi r^2) = \tau(2\pi rL) \tag{9-1}$$

where L is the length of the cylinder element, or

$$\tau = \frac{\Delta Pr}{2L} \tag{9-2}$$

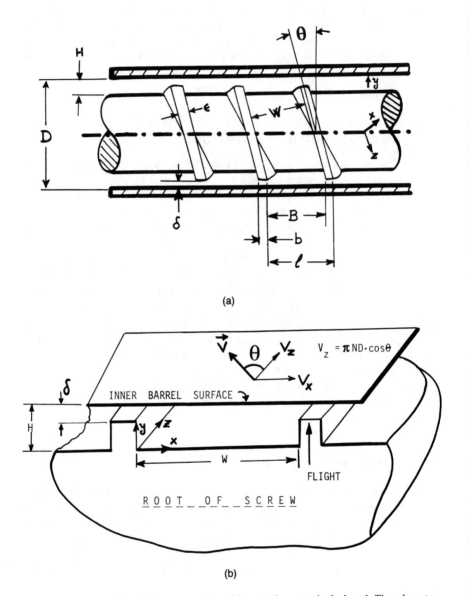

(a)

(b)

FIGURE 9-3.   (*a*) Simplified representation of the extruder screw in the barrel. The relevant geometric parameters are also represented. (Modified schematic of a diagram by Harper, 1981.) (*b*) Axes and particle velocity for flow in the extruder screw channel.

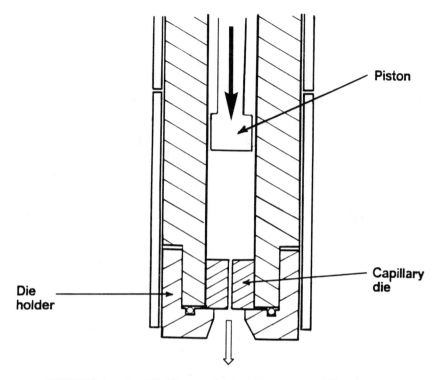

FIGURE 9-4.   Simplified diagram of a piston/plunger-type capillary rheometer.

At the wall, the shear stress has the value

$$\tau_{\text{wall}} = R \cdot \frac{\Delta P}{2L} \tag{9-3}$$

For a Newtonian fluid, from the definition of viscosity (eq. (1-6)), one has

$$\dot{\gamma} = -\frac{dv}{dr} = \frac{\tau}{\eta} \tag{9-4}$$

where $\dot{\gamma}$ is the shear rate and $v(r)$ is the fluid velocity in the cylinder element of radius r.

Since from equations (9-2) and (9-3), $\tau = \tau_{\text{wall}} \cdot r/R$, by substituting in equation (9-4) this expansion of τ yields the simple relation

$$\frac{dv(r)}{dr} = -r \cdot \frac{\tau_{\text{wall}}}{\eta R} \tag{9-5}$$

or

$$dv(r) = -\frac{\tau_{wall}}{\eta R} \cdot r \, dr \qquad (9-6)$$

Finite integration over the $(0, v)$ interval of velocities yields

$$v = \int_0^v dv = -\frac{\tau_{wall}}{\eta \cdot R} \int_R^0 r \, dr \qquad (9-7)$$

which is a parabola with the maximum velocity, $v_{max} = \tau_{wall} R/2\eta$, occurring on the axis of the capillary tube, for $r = 0.0$.

The average velocity, $\tilde{v}$, is also readily calculated in this case (Harper, 1981):

$$\tilde{v} = \frac{1}{s} \int v \, ds = \frac{\tau_{wall}}{R^3 \eta} \int (R^2 - r^2) \, dr \qquad (9-8)$$

or

$$\tilde{v} = \tau_{wall} \cdot \frac{R}{4\eta}$$

The normalized velocity ratio is therefore

$$\frac{v}{\tilde{v}} = 2\left[1.0 - \left(\frac{r}{R}\right)^2\right] \qquad (9-9)$$

A similar expression is obtained for a rectangular slit or die:

$$\frac{v}{\tilde{v}} = \frac{3}{2}\left[1.00 - \left(\frac{r}{W}\right)^2\right] \qquad (9-10)$$

where W is the width of a narrow slit opening and r is the distance between two parallel plates, with the slit opening between the plates.

RELATIONSHIP BETWEEN VISCOSITY AND FLOW RATE

From equations (9-4), (9-7), and (9-8) one can derive the relationship between the flow rate, Q, and viscosity in the capillary flow.

Since

$$\dot{\gamma} = -\frac{dv}{dr} = \frac{\tau}{\eta} \quad \text{and} \quad Q = \pi R^2 \cdot \tilde{v} \qquad (9\text{-}11)$$

one can relate Q to η through equations (9-9), (9-4), and (9-11). Thus, from equation (9-9), $v = 2\tilde{v} - 2\tilde{v} \cdot (r/R)^2$, and

$$\dot{\gamma} = -\frac{dv}{dr} = 4\tilde{v} \cdot \frac{r}{R^2} \qquad (9\text{-}12)$$

Substituting $\tilde{v}$ from equation (9-11) into equation (9-12), one obtains

$$\dot{\gamma} = 4Q \cdot \frac{r}{\pi R^4} \qquad (9\text{-}13)$$

On the other hand, one notes that the viscosity can be calculated from the shear stress and shear rate values at the wall:

$$\eta = \frac{\tau_{wall}}{\dot{\gamma}_{wall}} \qquad (9\text{-}14)$$

The shear rate at the wall is obtained from (eq. (9-13)) by setting $r = R$ and, therefore,

$$\dot{\gamma}_{wall} = \frac{4Q}{\pi R^3} \qquad (9\text{-}15)$$

Combining equation (9-15) with (9-14) and (9-3), one obtains

$$\eta = \frac{(R \cdot \Delta P/2L)}{(4Q/\pi R^3)}$$

or

$$\eta = \pi \cdot \Delta P \cdot R^4/8QL \qquad (9\text{-}16)$$

Equation (9-16) is the *Hagen–Poiseuille law* of laminar flow, and it allows one to calculate the viscosity of a Newtonian fluid from the capillary flow rate, pressure gradient, and geometric parameters (R and L) of the capillary.

APPARENT VISCOSITY OF
NON-NEWTONIAN FLUIDS

There have been attempts to expand the applicability of such calculations to non-Newtonian, isothermal flows in a capillary rheometer. In non-Newtonian fluids, however, the ratio of shear stress to shear rate varies across the capillary; this ratio, $\tau/\dot{\gamma}$, is the apparent viscosity, η_{app}. At the wall, the shear stress can be calculated as $R \cdot \Delta P/2L$ (eq. (9-3)); therefore, it would be possible, in principle, to calculate η_{app} at the wall, if $\dot{\gamma}_{wall}$ was known for a non-Newtonian fluid. Rabinowitsch proposed a calculation of $\dot{\gamma}_{wall}$ from the *average* flow rate for a non-Newtonian fluid for which the dependence $v = v(r)$ is known (Harper, 1981):

$$\overline{Q} = \int 2\pi r \cdot v(r) \cdot dr = \frac{\pi r^2 v(r)}{\left[-\pi \cdot \int r^2 \, (dv/dr) \, dr \right]} \tag{9-17}$$

Assuming that there is no slip at the wall (or a lubricated material), since $v = 0$ at the wall (for $r = R$), one has that

$$Q = -\pi \int r^2 v'(r) \, dr \tag{9-18}$$

From equations (9-2) and (9-3), on the other hand, one has that $\tau/\tau_{wall} = r/R$, and one can therefore substitute an expression of τ instead of r in equation (9-18), so that

$$Q = -\lambda \int \left(\frac{\tau}{\tau_{wall}} \right)^2 R^2 \cdot v'(\tau) \, \frac{R}{\tau_{wall}} \cdot d\tau$$

or

$$Q = \frac{\pi R^3}{\tau_{wall}^3} \int \tau^2 v'(\tau) \, d\tau \tag{9-19}$$

Substituting this expression of Q in equation (9-15) yields an expression for calculating $\dot{\gamma}_{wall}$, and therefore η_{app} at the wall:

$$\dot{\gamma}_{wall} = \frac{4Q}{\pi R^3} = -\frac{4}{\tau_{wall}^3} \int \tau^2 v'(\tau) \, d\tau$$

and

$$\eta_{app}^w = \frac{\tau_{wall}}{\dot{\gamma}_{wall}} = \frac{\tau_{wall}}{(-4/\tau_{wall}^3) \cdot \int \tau^2 v(\tau) \cdot d\tau}$$

or

$$\eta_{app}^{w} = -\frac{(\tau_{wall}^{4}/4)}{\int \tau^{2} v'(\tau) \cdot d\tau}$$

Furthermore, since $\tau_{wall} = R\Delta P/2L$, one obtains

$$\eta_{app}^{w} = \left(\frac{R\Delta P}{L}\right)^{4} 64 \int \tau^{2} \cdot v'(\tau) \cdot d\tau \qquad (9\text{-}20)$$

Equation (9-20) would allow one to calculate η_{app} at the wall if the dependence of the velocity on r or τ was known. For specific types of flows encountered in foods, such as Bingham plastic or pseudoplastic flows (Fig. 1-4), one can therefore estimate the apparent viscosity at the wall of the capillary rheometer from equation (9-20) by measuring either the average flow rate or the pressure difference between the two ends of the tube.

END CORRECTIONS

At both the entrance and the end of the capillary tube, pressure drops occur, thereby reducing the pressure, ΔP, in equation (9-2). Corrections for the expansion losses of pressure at the exit of the tube were developed by Bagley (1954); the constriction losses at the entrance of the capillary tube can be reduced by using a gradually tapering of the tube entrance. For a short capillary, Bagley (1954) proposed that the shear stress at the wall can be calculated as:

$$\tau_{wall} = \frac{\Delta P}{2(L/r + L_{e}^{1}/R)} \qquad (9\text{-}21)$$

where τ_{wall} is the shear stress at the wall (in N/m^{2}), ΔP is the pressure drop (in N/m^{2}), L is the length of the capillary, L_{e}^{1} is the equivalent length of the capillary that would increase the pressure difference, ΔP, to equal the pressure drop caused by the end effects, and R is the radius of the capillary. Rogers (1970) proposed a practical procedure for determining the end correction term (L_{e}^{1}/R) under isothermal extrusion conditions. One would employ several short capillary dies with various L/R ratios, and then plot (log $\dot{\gamma}_{wall}^{a}$) against (log ΔP), with $\dot{\gamma}_{wall}^{a} = 4Q/\pi R^{3}$ (from eq. (9-15)), which should give a straight line for each die. Next, one would plot ΔP against L/R for different values of $\dot{\gamma}_{wall}^{a}$ and then one extrapolates each line to $\Delta P = 0$, which yields a value of L_{e}^{1}/R against $\dot{\gamma}_{wall}^{a}$; one should thus obtain a straight line from which specific values of L_{e}^{1}/R could be obtained by interpolation, or extrapolation, at the value of the $\dot{\gamma}^a_{wall}$ that is selected in the experiment or for prediction. By employing these values of L^1_e/R in equation (9-21), one can calculate the corrected values of τ_{wall} (τ^{corr}_{wall}) against (log $\dot{\gamma}^a_{wall}$). These values would give a straight line, with the slope, b (or n), the *flow behavior index* (see also the following section on extrusion of pseudoplastic materials). Several numerical examples of such calculations were given by Harper (1981, pp. 38–42). However, these corrections do *not* take into account the effects of *turbulent* flow, when it occurs.

EXTRUSION OF BINGHAM PLASTIC MATERIALS UNDER ISOTHERMAL CONDITIONS

Organic polymers usually have Bingham plastic behavior (Fig. 1-4), as do a few food materials, such as corn flour doughs in the range of 40% to 50% moisture content (w/w). The *Bingham viscosity*, η_B, is defined as

$$\eta_B = \frac{(\tau - \tau_0)}{\dot{\gamma}} \tag{9-22}$$

and the flow equations remain relatively simple, the only complication being the *yield parameter*, τ_0. Since η_B and τ_0 are independent characteristics of a Bingham plastic material, both need to be determined to completely specify the rheological behavior of the material. Equations (9-2) through (9-16) can be readily modified to include the effect of τ_0. Thus,

$$\tau^B_{wall} - \tau_0 = \frac{R\Delta P}{2L} \tag{9-23}$$

and the velocity profile is given by the equation

$$v = \frac{\tau_{wall} - \tau_0}{2\eta_B \cdot R} \cdot (R^2 - r^2) \tag{9-24}$$

whereas the normalized velocity ratio has the same form as in Newtonian flow, and equation (9-13) is still valid. On the other hand,

$$\eta_B = \frac{\tau_{wall} - \tau_0}{\dot{\gamma}^a_{wall}} \tag{9-25}$$

and $\dot{\gamma}^a_{wall} = 4Q/\pi R^3$. Therefore, one has

$$\eta_B = \frac{R \cdot \Delta P/2L - \tau_0}{4Q/\pi R^3}$$

or

$$\eta_B = (\pi \cdot \Delta P \cdot R^4 / 8QL) - \pi R^3 \cdot \tau_0 / 4Q \qquad (9\text{-}26)$$

which is slightly more complicated than the corresponding expression (eq. (9-16)) for Newtonian, isothermal flow. The end corrections (eq. (9-21)), of course, still apply, and one has to calculate the shear stress at the wall as

$$\tau_{\text{wall}} = \frac{\Delta P}{2(L/R + L_e^1/R)} \qquad (9\text{-}27)$$

EXTRUSION OF PSEUDOPLASTIC FOOD DOUGHS UNDER ISOTHERMAL CONDITIONS

Pseudoplastic and *dilatant* flow behaviors are encountered in many food doughs. An approximate treatment of pseudoplastic or dilatant flow is often attempted with the *power law*:

$$\tau = a\dot{\gamma}^b \qquad (9\text{-}28)$$

where a is the dough *consistency coefficient* and b is the *flow behavior index*. (In extrusion technology the notation n for a and m for b is often used.) For pseudoplastic fluids, $b < 1$, whereas for dilatant flow, $b > 1$. (The Newtonian fluid case corresponds to $b = 1.00$ and $a \equiv \eta$.) A yield stress, τ_0, term can be added to the right-hand side of this equation in order to treat non-Bingham plastic flows (see Fig. 1-4 for all of these flow behaviors). The latter form of equation (9-28) is known as the *Herschley–Buckley model*. Equation (9-28) approximates reasonably well the behavior of pseudoplastic materials that are in the middle range of the shear rate ($\dot{\gamma}$) values only. The *Reiner–Philippoff model*

$$\tau = -\frac{\eta_0 + 1 + (\tau/\tau_s)^2}{(\eta_0 - \eta_\infty - \dot{\gamma})} \qquad (9\text{-}29)$$

seems to fit the data for many pseudoplastic materials, especially at *low* and *high* shear rates, where equation (9-28) does not fit the data very well. However, equation (9-29) increases the number of parameters required for the fitting, which is not generally desirable.

The calculation of the velocity profile for power-law fluids can be done in a manner analogous to the derivation in equations (9-5) and (9-9); thus, the normalized velocity ratio is calculated to be

$$v/\bar{v} = \frac{3b + 1}{b + 1} \cdot \left[1 - \left(\frac{r}{R}\right)^{b+1/b}\right] \qquad (9\text{-}30)$$

for power law, pseudoplastic fluids. Note that the value of v_{max}/v is $(3b + 1)/(b + 1)$, which is less than 2.0 if $b < 1$, the value obtained for Newtonian flow. When $b = 0$ one would obtain $v(r) = \bar{v}$, that is, *plug flow*. The normalized velocity profile becomes flatter as b decreases, or as the material flow has stronger pseudoplastic character. Note also that neither the viscosity η nor a, the consistency coefficient, have any effect on the velocity profile. However, η is inversely related to flow rate, as shown in equation (9-16). For power-law fluids flowing through a narrow, rectangular slit of width w, one can readily show that

$$\dot{\gamma}_{wall}^a = \frac{2b + 1}{3b} \cdot \frac{3Q}{2T^2 \cdot w} \qquad (9\text{-}31)$$

(Harper, 1981), where T is half the slit thickness and w is the slit width. Since $\tau_{wall} = a\dot{\gamma}_{wall}^b$, substitution in the previous equation yields

$$a = \frac{\tau_{wall}}{[(2b + 1/3b) \cdot (3Q/2T^2) \cdot w]^b} \qquad (9\text{-}32)$$

which shows that the flow rate raised to the power b and the consistency coefficient a are inversely related for a power-law fluid. The relationship between the flow parameters (that is, a and b) and flow rate, Q, is no longer linear as it was in the case of Newtonian fluids (eq. (9-16)). Before applying the end corrections one still has for capillary flow that $\tau_{wall} = R \cdot \Delta P/2L$, and therefore $a = (R \cdot \Delta P/2L)/[(2b + 1/3b) \cdot (3Q/2T^2) \cdot w]^2$

$$Q = 2T^2 \cdot w \cdot (b/2b + 1) \cdot (a \cdot 2L/R\Delta P)^{1/b} \qquad (9\text{-}33)$$

which shows that, unlike the case of Newtonian flow where Q is proportional to ΔP (eq. (9-16)), the flow rate is no longer proportional to the pressure difference for pseudoplastic flow.

ISOTHERMAL NEWTONIAN FLOW IN A SINGLE-SCREW EXTRUDER

The piston-type capillary rheometer simplifies the flow calculation for the extrusion of Newtonian, or non-Newtonian, fluids under isothermal conditions.

The results for the flow in a single-screw extruder are suggestive, but the presence of a flighted screw instead of a piston does complicate the flow calculations significantly.

Here the effect of the screw on the flow in an extruder is considered by assuming an incompressible fluid for which the *Navier–Stokes hydrodynamic flow equations* would be applicable:

$$\rho \cdot (\partial v_{x_i}/\partial t) + \vec{v} \cdot (\nabla \vec{v}) = -\frac{\partial P}{\partial x_i} + \eta \cdot \Delta v_{x_i} \qquad (9\text{-}34)$$

where $i = 1, 2, 3$, and $x_1 \equiv x$, $x_2 \equiv y$, $x_3 \equiv z$, ρ is the density, η is the viscosity of the fluid, and $\vec{v}$ is the *fluid velocity*. (One should also consider sedimentation effects caused by gravity to be negligible under extrusion conditions.)

The configuration of the extruder screw is schematically represented in Figure 9-3b, with the extruder barrel axis in the z-direction, or the direction of flow, and the y-axis as the flight axis, from the root to the top of the screw. The calculations are easier if these axes are stationary and the fluid moves around the screw, as if the barrel were rotating instead of the screw. Since motion is relative, the results of the calculation are equivalent for the two situations, and this computational simplification is both justified and convenient. Neglecting *tangential* flows in the x- and y-directions implies that $\partial v_x/\partial x + \partial v_y/\partial y = 0$, and the extruder output, Q, can be calculated by considering the flow only in the z-direction (i.e., parallel to the screw or extruder, long axis). For a number p of channels that are parallel, the volumetric extruder output is

$$Q_V = p \int \int v_z \cdot dy \, dx \qquad (9\text{-}35)$$

where H is the *height* and W is the *width* of the flight of the extruder screw (Harper, 1981). The double integral creates two flow components, the first of which is proportional to the velocity at the top of the flight, $V_z = v_z (x, H)$ (assuming there is no slip between screw channels), and corresponds to the *drag flow* (Q_d) resulting from viscous drag, and is proportional to the rotational speed, N, of the screw. The second flow component is the *pressure flow* (Q_p), and is proportional to the pressure gradient, $\partial P/\partial z$, along the screw channel (in the z-direction). Under normal extrusion conditions, the two flows oppose each other (for Newtonian fluids), and $\partial P/\partial z$ is negative because the pressure is highest at the discharge. The net volumetric extruder output is therefore

$$Q_V = Q_d + Q_p \qquad (9\text{-}36)$$

with $Q_d \sim p \cdot V_z \cdot WH$, $Q_p \sim pWH^3 \cdot (\partial P/\partial z)$, and $V_z = \pi \cdot ND \cdot \cos \Theta$; D is the screw diameter and Θ is the screw inclination angle between $\vec{v}$ (fluid velocity) and the z-axis of flow.

In food extrusion, it sometimes happens that $Q_d \gg Q_p$, and since $Q_d \sim V_z \sim N$, the net volumetric output of the extruder, Q_V, will also be approximately proportional to the rotation speed of the extruder (single) screw. Various screw shape correction factors are needed for both *drag* and *pressure* flow calculations of extruder output in order to account for flight "edge" effects, that is, the channel wall effects on the flow that increase as the ratio H/W increases. Harper (1981; pp. 51 and 52) gives analytical expressions of such shape factors for Newtonian flow. When $H/W = 0$, these shape, or edge, correction factors become, of course, equal to 1.00 (i.e., *no correction*, as expected from the piston-capillary rheometer calculations). In addition, the calculated extruder output also needs to be corrected for the extruder end effects discussed in the previous section, under *end corrections* for capillary rheometers (eq. (9-21)).

VELOCITY PROFILE FOR NEWTONIAN FLOW IN A SINGLE-SCREW EXTRUDER

The calculation of the velocity profile in the screw channel proceeds by applying the *Navier–Stokes equations* (eq. (9-34)) to the configuration in Figure 9-3, and assuming there is no slip in the channel or at the inner barrel surface under steady-state extrusion conditions; that is, $\partial v_{x_i}/\partial t = 0.0$ for $i = 1, 2,$ and 3. Therefore,

$$\eta \left(\frac{\partial^2 V_z}{\partial x^2} + \frac{\partial^2 v_z}{\partial y^2} \right) = \frac{\partial P}{\partial z} \tag{9-37}$$

Integrating this equation twice gives

$$v_z = \frac{y^2 - Hy}{2\eta} \cdot \frac{\partial P}{\partial z} + V_z \cdot \frac{y}{H}$$

or

$$v_z = \left(\frac{Q_p}{pWH^3} \right) \frac{(y^2 - Hy)}{(2\eta + Q_d \cdot y)/pWH^2} \tag{9-38}$$

which can be rearranged as

$$v_z = V_z \left[(1 - 3a) \left(\frac{y}{H} \right) + 3a \left(\frac{y}{H} \right)^2 \right] \tag{9-39}$$

(Harper, 1981), with $a = Q_p/Q_d$.

The velocity profile in the down-channel z-direction, which is represented by this equation, was plotted by Paton et al. (1970) and is illustrated in Figure 9-5a. In Figure 9-5a one can see the two superimposed flow components, the drag and pressure flows for the Newtonian fluid, as well as their *separate* velocity profiles (in the top part of Fig. 9-5a). Note that only drag flow is present at the *open* discharges, as expected, while the two flow components are combined at the *closed* discharge. Across the screw channel (in the x-direction), the flow and cross-channel velocity profiles, v_x, were also computed by Paton et al. (1970). The results show that there is no flow in the x-direction ($\int v_x \cdot dx = 0.0$), and that

$$v_x = \frac{y}{H} \frac{2 - 3y}{H} V_x \qquad (9\text{-}40)$$

(Harper, 1981), which shows that the cross-channel velocity is *independent* of the pressure flow. The actual, or resultant, movement of a particle in the screw channel is obtained by considering that $\vec{v} = v_x \cdot \vec{i} + v_y \cdot \vec{j} + v_z \cdot \vec{k}$ and that $v_y = 0$. The resultant particle paths in the screw channel are therefore *closed helical paths* from the vector sum of $v_x \cdot \vec{i}$, with $v_z \cdot \vec{k}$. A particle moves across the channel near the screw root and then comes out at the top of the channel when it reaches a flight. These velocity profiles also need to be corrected for edge effects, as discussed earlier.

In a food extruder, the flow in the barrel sections is coupled to the flow through the die, at the extruder discharge. For circular and slit-shaped dies, the velocity profiles for Newtonian flow would be determined by equations (9-9) or (9-10) and (9-11), respectively, to which one would make the end corrections (eq. (9-21)). The flow through the die is obtained with the Hagen–Poiseuille equation (eq. (9-16)), which is considerably simpler than the equations for the flow in the screw channels (eqs. (9-39) and (9-40)). Another problem is caused by the temperature and fluid-viscosity differences between the fluid in the die and the fluid in the screw channels. The problem can be addressed experimentally by measuring the flow properties at the die discharge with a mounted (thermostated) capillary rheometer, as discussed in the next section. If the fluid viscosity at the die is η_d and that in the screw channels is η_c, it can be shown (Harper, 1981) that the total volumetric flow of an extruder under steady, Newtonian flow conditions is

$$Q_V \sim \frac{G_1 N}{1 + (\eta_d/\eta_c) \cdot (G_2/L)}, \qquad (9\text{-}41)$$

where G_1 and G_2 are parameters that depend only on the geometrical characteristics of the extruder screw and flight, and L is the length of the barrel. The

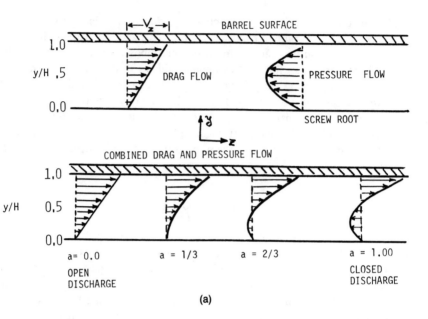

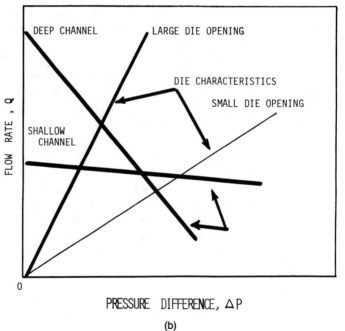

FIGURE 9-5.   Selection of the operating conditions of the extruder: determination of the operation points (modified from Harper, 1981).

pressure difference is then

$$\Delta P \sim G_1 \cdot \frac{N}{K/\eta_d + G_2/\eta_c \cdot L} \tag{9-42}$$

where K is a geometric parameter depending on the die opening. Appropriate shape factor corrections need to be applied in equations (9-41) and (9-42) to obtain the correct extruder output. The relationship between the volumetric flow output through the die, Q_V^D, and the pressure drop, ΔP, is remarkably simple in this case and, of course, resembles equation (9-16):

$$Q_V^D = \frac{K\Delta P}{\eta_D} \tag{9-43}$$

This simple equation indicates that the operating point of the extruder (Fig. 9-5b) is at the crossing between the die characteristic curve and the screw characteristic curve. Changing from a large to a small die can result in a shallow flighted screw having a larger output than a deep flighted screw. Although equation (9-16) indicates that a decrease in viscosity as a result of increasing temperature will increase the flow rate, in fact only the pressure flow component increases in the extruder; the volumetric output Q_V^C through the screw channels decreases when the pressure flow component increases. On the other hand, the flow at the die, Q_V^D, increases for a given difference in pressure, compensating for the decreased Q_V^C. Therefore, the *net* extruder output may be little affected by increasing the temperature and decreasing the fluid viscosity in the barrel. Furthermore, *leakage flows* that occur as a result of the circular motion of particles within the channel also reduce the drag flow, and may cause a significant decrease in the extruder output.

ISOTHERMAL NON-NEWTONIAN FLOW IN SINGLE-SCREW EXTRUDERS

The separation of the total volumetric flow into *drag* and *pressure* flow components (eq. (9-36)) *does not hold for non-Newtonian fluids* since there is no longer a uniform viscosity defined for a non-Newtonian fluid such as a food dough.

Various regions of the fluid in the extruder screw channels experience different shear rates and shear stresses, and there is no linear velocity gradient (Fig. 1-3) through the fluid for an applied shear stress, as was the case for Newtonian fluids. For narrow dies one still may be able to define an *effective* shear rate and carry out an approximate computation of the flow through the die if the rheological parameters of the food dough or fluid, such as the *a* and *b* values of power-law fluids in the die, were known. For a short die, the end-correction factor, L'_e/R, is about 5, but can be determined more accurately with dies of various lengths, as described under end corrections.

TWO EXAMPLES OF CAPILLARY RHEOMETRY APPLICATIONS

Extrusion cooking has the major advantage of making the processing of very large quantities of food materials possible in a continuous high-temperature, *short-time* (HTST) mode. As discussed at the beginning of this chapter, the twin-screw extruders have certain practical advantages in operation, but they are harder to analyze, or to predict, in terms of the operating conditions for new formulations. Recently, attempts have been made to determine several important operating parameters of twin-screw extruders by joining a capillary rheometer to the die of the extruder. Two recent studies of the extrusion cooking of starches illustrate the application of the theoretical analyses of extrusion presented briefly in the previous sections. Unlike single-screw extruders, however, twin-screw extruders are fed either volumetrically or gravimetrically, which means that the degree of filling of the food zone is less than 100%. As a result, the extruder output through the die can be varied independently of the twin-screw speed, so that the relations developed previously for the net extruder output of a single-screw extruder (for example, eq. (9-16)) are no longer applicable.

Capillary Rheometry of Maize (Corn) Starch Extrusion with a Twin-Screw Extruder

In a recent report by Parker et al. (1990), a Baker-Perkins MPF50D corotating twin-screw extruder was employed for extrusion cooking of hydrated corn (maize) starch. An extruder-fed cylindrical rheometer, as well as a capillary rheometer, were then employed to determine the flow characteristics of the extruded material. These characteristics were studied as a function of the extruder barrel temperature, the water content of the extruded material, and specific mechanical energy (SME) input, in order to determine whether breakdown, or degradation, of corn starch occurred in the twin-screw extruder at the higher shear rates. Previously, Vergnes and Villemaire (1987) showed that the extrusion of low-moisture molten maize starch caused the degradation of the amylopectin (and/or amylose) components of starch that depends upon the specific mechanical energy input.

Schematic diagrams of the capillary rheometer and the extruder-fed cylindrical rheometer are shown in Figures 9-6 and 9-7, respectively. The rheometry with the extruder-fed instrument was carried out either at constant rheometer

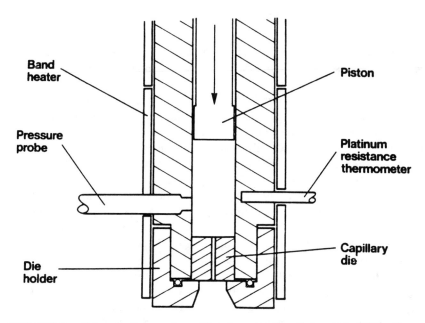

FIGURE 9-6. Schematic diagram of a capillary rheometer (from Parker et al., 1990, with permission).

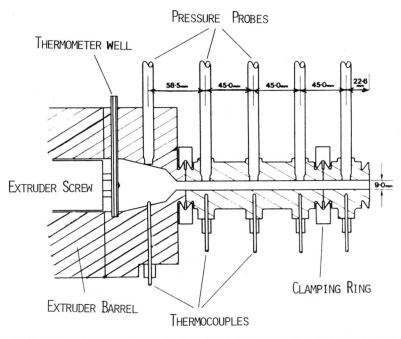

FIGURE 9-7. Diagram of the extruder-fed cylindrical rheometer (from Parker et al., 1990, with permission).

length or at constant water content. The variations of the apparent viscosity with shear rate, set temperature, rheometer length, and water content were measured.

End corrections were determined by the Bagley method, as described earlier in this chapter (eq. (9-41) and related discussion). Plots of the pressure drop, Δp, versus the length/diameter ratio, L/D, of the capillary rheometer are shown in Figure 9-8 for drum-dried maize starch at various $\dot{\gamma}_{wall}(s^{-1})$ shear rates. Note that the extrapolated linear regression lines had intercepts that were as large as the overall pressure drop, that is, the end pressure corrections of the capillaries were quite substantial. The *apparent* shear viscosity of drum-dried maize starch measured with the capillary rheometer at a wall shear rate, $\gamma_{wall} = 100 \text{ s}^{-1}$, decreased with increasing temperature between 100° and 150°C, as was expected for viscous flow. Shear thinning was measured for native maize starch by extruder-fed rheometry, and is reproduced (with permission) in Figure 9-9. The apparent shear viscosity, $\eta_f^{app} = \tau_w/\dot{\gamma}_{wall}^{app}$ (Boydson, 1981), is a function of the shear rate between 40 and 180 s^{-1} for a water content m_w of 0.21 (g/g; w/w), at set barrel temperatures of 100°, 120°, and 150°C. The value of η_f^{app} decreased when the temperature increased from 100° to 120°, but it did not decrease significantly when the set barrel temperature increased from 120° to 150°C at $m_w = 0.21$. At $m_w = 0.28$ and $m_w = 0.38$, however, the shear-thinning was apparent only up to $\gamma_{wall} \sim 100 \text{ s}^{-1}$, above which the apparent

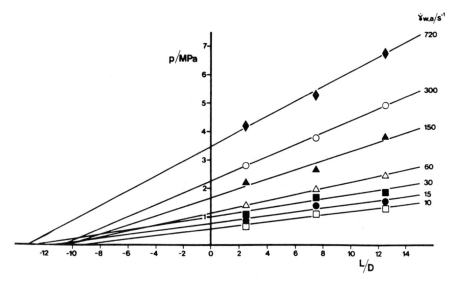

FIGURE 9-8.   Pressure drop, Δp, versus length/diameter ratio, L/D, for drum-dried starch at a temperature of 130°C and a water content of 0.28 (from Parker et al., 1990, with permission).

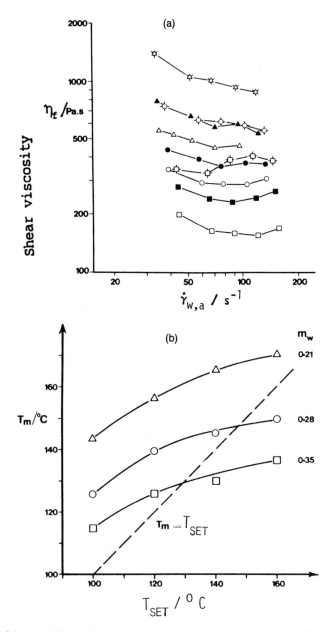

FIGURE 9-9. (a) Shear viscosity, η_f, as a function of shear rate, $\dot{\gamma}_{w,a}$, for different set barrel temperatures, T_{set}, and water content, m_w, measured using the cylindrical rheometer. m_w:0.21, T_{set}; ⚹, 100°C; ▲, 120°C; △, 160°C; m_w: 0.28, T_{set}: ✿, 100°C; ●, 120°C; ○, 160°C; m_w: 0.35, T_{set}: ⬡, 100°C; ■, 120°C, □, 160°C (from Parker et al., 1990, with permission). (b) Melt temperature, T_m, as a function of set barrel temperature, T_{set}, for different water contents: △, 0.21; ○, 0.28; □, 0.35. Solid feed rate = 20 kg hr^{-1} (from Parker et al., 1990, with permission).

viscosity increased with γ_{wall}. Overall, the apparent shear viscosity decreased with increasing water content, m_w, at all temperatures, except for $m_w = 0.35$ at $T_{set} = 100°C$ and $\gamma_{wall} > 70$ s^{-1}. This apparent paradox was explained by noting that the melt temperature, T_m, was not proportional to the set barrel temperature, and increased less with T_{set} at $m_w = 0.35$ than at $m_w = 0.21$. Because of the short residence time of the hydrated maize starch in the extruder there was not enough time for the melt and barrel temperatures to equilibrate. Note that in such a case the *isothermal* operating conditions assumed in the previous theoretical analysis are not being met, and therefore one cannot use the methodology described for isothermal extrusion processes to carry out a precise analysis of the results described in this section. Other complications are caused by the mechanical degradation of the maize starch produced by the twin-screw extruder at the higher shear rates (rotation speeds) of the *specific mechanical energy input*. The latter represents the mechanical work dissipated per unit mass of net extruder output, or throughput.

Power law indices a and b were also determined from the rheometry data for hydrated maize starch, with $m_w = 0.21$, 0.28, and 0.35. The results are summarized in Table 9-1. Interestingly, the extruder-fed rheometer showed a higher value for the power law index, b, at $m_w = 0.35$ than did the other two instruments. As expected, the b value increased with m_w. At higher water contents ($m_w \geq 0.35$), the degradation of the macromolecular components of starch was significantly different from that of the lower m_w region, which agreed with the results presented in the next section for wheat starch extrusion cooking. A comparison of the corrected, apparent viscosities, η, which were determined with three rheometers as a function of the shear rate γ_{wall} at $m_w = 0.28$ and 130°C, is reproduced (with permission) in Figure 9-10. The overall agreement was quite reasonable, although it could not be claimed as systematic. Interestingly, the rheoplast measured values overlapped with the other rheometers only for SME $= 600$ $\delta \cdot$ g^{-1}. However, all log η_{app} versus log γ_{wall} plots were linear for the range of γ_{wall} values between 40 and 190 s^{-1}. The cylindrical rheometer could also be employed to determine the entrance and exit pressure differences as a function of shear stress (Fig. 9-11). Both pressures appear to vary almost

TABLE 9-1   The Coefficients in Equations (9-1) and (9-2) Obtained by Regression of the Cylindrical Rheometer Data

m	T_0 (°C)	W_0 (J · g^{-1})	K (Pa · s^n)	n	T Range (°C)	W Range (J · g^{-1})
0.21	160	600	9270	0.31	142–170	310–1020
0.28	140	350	3280	0.48	126–151	190–630
0.35	125	250	890	0.68	107–136	160–340

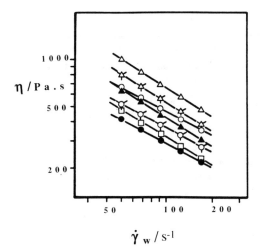

η / P a . s

$\dot{\gamma}_w$ / s^{-1}

FIGURE 9-10. Shear viscosity, η_f, as a function of shear rate, $\dot{\gamma}_w$, obtained using different rheometers at a water content of 0.28 and a temperature of 130°C. Capillary, $\square$: extruder-rod cylindrical, $\bigcirc$, $\female$, $\bullet$; Rheoplast, $\triangle$, $\star\!\!\star$, $\blacktriangle$, at SME of 200 Jg^{-1} and 600 Jg^{-1}, respectively (from Parker et al., 1990, with permission).

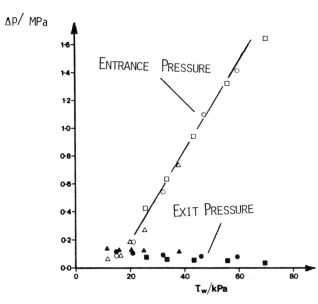

FIGURE 9-11. Entrance and exit pressure, Δp, as a function of shear stress for the 189-mm-long cylindrical rheometer at water content 0.28 and set temperatures: $\square$, $\blacksquare$, 100°C, $\bigcirc$, $\bullet$, 120°C; $\triangle$, $\blacktriangle$, 160°C (from Parker et al., 1990, with permission).

linearly in the middle range of τ_{wall} values of 25 to 65 kPa and $m_w = 0.28$, but the entrance temperature appears to follow equation (9-3), whereas the exit pressure does not. Note that the pressure difference dependence on τ_{wall} is independent of temperature or shear rate. Similar behavior has been previously reported for synthetic polymer (plastic) melts (Han, 1971). Water appears to play a dual role in food dough extrusion: (1) as a plasticizing agent, that is, water acts to increase the expansion of the food melt after discharge at the die, and (2) the lower bulk density extrudates are formed from the lower viscosity melts.

A more detailed study of the degradation of starch caused by the twin-screw extruder was recently carried out by van Lengerich (1990) and is discussed next.

Physical and Chemical Analysis of Extrudates in Relation to the Rheological Behavior of Hydrated Wheat Starch

Both heat and mechanical work increase the internal energy of the food dough being extruded and this causes it to be *plasticized*. Sudden water evaporation at the discharge of the extruder causes the plasticized food dough to expand rapidly and acquire a characteristic shape and texture. Among the processes induced by the increased internal energy of the food dough system are starch gelatinization, complex formation between lipids and starch, protein denaturation/coagulation, Maillard reactions, and several other types of chemical reactions activated within the extruder. Such reactions inside the extruder depend on both the composition of the raw material and the operating parameters of the extruder. Because of this complexity food extrusion technology is semiempirical at best. The scientific literature on food extrusion, on the other hand, is mainly concerned with the physical and chemical properties of the extruded products (*extrudates*), as pointed out by van Lengerich (1990). The latter properties are, of course, determined by the specific extrusion conditions existing in the various types of extruder now in use. The disadvantages of the empirical approach are many and are costly for the food industry: for example, the results obtained with one extruder type cannot be readily transferred, or scaled up, to another; the calibration process is very costly and time-consuming; and the selected process may not be optimal for the food applications that are of immediate practical interest in a fast-changing, competitive market situation. The only advantage of the empirical approach is simplicity: anyone can empirically test parameters and list the results without regard for the underlying principles of operation.

It is therefore necessary to build predictive models that account for the thermodynamic variables of state, such as the internal energy and temperature (see also Chap. 2 on thermodynamics), as well as the time dependence/shear history of the product in the extruder, and to correlate such parameters with the structural and functional properties of the extrudates. Several techniques therefore need to be applied to characterize the numerous physical and chemical processes occurring during the food extrusion process. For example, it is essential to know the rheological properties of the plasticized food dough in order to predict the pressure building in production-scale extruders, to do a reliable scaleup of laboratory-scale extruders, the puffing behavior of the dough, or the net volumetric output of the extruder. On the other hand, the phase transformations of the hydrated food dough at various stages of extrusion can be studied by DSC or DTA techniques (see also the relevant sections on these techniques in Chap. 5), and the heat effects can be correlated with flow/mechanical properties by DTMA. In addition, structural or degradation changes caused by extrusion can be studied by X-ray diffraction, NMR, ESR, FT-IR, and several other established techniques for structure determination and conformation analyses. Fluorescence methods are often used to investigate the residence-time distribution of extruded particles in the food dough. Polarized light microscopy and SEM are also used to determine changes in the starch structure caused by extrusion. NMR techniques (especially ^{17}O NMR; Baianu et al., 1990; Mora-Gutierrez and Baianu, 1989; Kakalis and Baianu, 1988) are particularly suitable for hydration studies of food biopolymers and starch (Yakubu, Baianu, and Orr, 1990, 1991), as discussed in more detail in Chapter 6.

An attempt was recently made to employ several techniques and rheological measurements to determine the effects of the extrusion of hydrated wheat starch on the structure and molecular weight distribution of the starch polysaccharides (van Lengerich, 1990).

The starch granules are compressed, deformed, and begin to gelatinize in the feed and metering sections of an extruder (Fig. 9-1), if the water content is sufficiently high. As a result, the energy dissipation through viscous flow increases at this point, and the pressure buildup causes *leakage* flow. It seems that the viscous dissipation of mechanical energy leads to the degradation or breakdown of starch polysaccharides, especially amylopectin (van Lengerich, 1990). Structural changes in corn amylopectin and waxy corn starch during gelatinization were also recently reported (Mora-Gutierrez and Baianu, 1991) in ^{13}C NMR studies of hydrated corn starch powders and suspensions. These results were also compared with wheat and potato starches (Yakubu et al., 1991). As a result, significant *quantitative* differences were found between the water plasticizing effects of the three types of starch, although, *qualitatively* the behaviors were similar.

In the extruder, gelatinization of starch begins in the shear zone caused by the heat generated by friction. This process plus the external heating of the extruder barrel cause the starch granules to swell in the presence of sufficient water ($\geq 30\%$ to 35%, depending on the starch type). Water molecules then surround amylose or amylopectin molecules in the starch structure, increasing their mobility and breaking hydrogen bonds between the polysaccharide components. At the same time, many weak hydrogen bonds are formed between water molecules and exposed hydroxyl groups of the glucose units (see also the carbohydrate section in Chapter 4). When the starch leaves the shear (or melting) zone, it is often completely gelatinized; that is, the starch granules do not show any birefringence when illuminated by polarized light under the microscope, and they are largely disrupted. In such starch extrudates, the *enthalpy* required to complete gelatinization, as measured by DSC, is negligible or zero. Furthermore, an enzyme hydrolyzability test of the extruded starch also indicates that the gelatinization is complete (van Lengerich, 1990). The appearance of the compressed hydrated starch at the end of the shear zone is that of a plastic, continuous dough. In the die section the starch dough, or melt, is compressed even further. Upon discharge, moisture is flushed out of the dough (it evaporates), and the starch material suddenly expands and puffs. Undoubtedly, at this point several *rapid irreversible* processes and phase transformations occur that determine the shape and texture of the extrudate.

The effects of the internal energy of the hydrated starch system inside the extruder were recently measured by using the concept of specific mechanical energy (SME) input (Meuser and van Lengerich, 1984), defined as

$$\text{SME} = M_d \cdot \frac{\omega}{\dot{m}} \qquad (9\text{-}44)$$

where M_d is the applied torque on the screw in $N \cdot m$, ω is the angular velocity of the extruder in s^{-1}, and $\omega = dm/dt$ is the net mass output or throughput of the extruder in kg/h. The studies reported by van Lengerich employed a Werner-Pfeleiderer twin-screw extruder (model Continua 37) with an L/D ratio of 12. The flow curves of the hydrated wheat starch extrudates were measured with a *slit* rheometer that was joined to the extruder's die plate through a transmission piece. The l/h ratio of the slit die was 150. The apparent shear rate was calculated with the equation

$$\dot{\gamma}_{\text{app}} = 3\dot{Q}/2B \cdot h^2 \qquad (9\text{-}45)$$

where $\dot{Q} = dQ/dt$ is the volumetric output of the extruder, B is the slit width, and h is half of the slit height. Such $\dot{\gamma}_{\text{app}}$ values were corrected by the Rabinowitsch method (Rabinowitsch, 1929):

$$\dot{\gamma}_{\text{corr}} = \frac{(2b + 1) \cdot H}{b} \cdot \frac{3\dot{Q}}{2B} \cdot h^2 \qquad (9\text{-}46)$$

where the flow index, b (often labeled as n), is defined as $b = \partial(ln\dot{\tau})/\partial(ln\dot{\gamma}_{app})$. The shear stress, τ (presumably at the wall of the slit rheometer), was calculated with the equation

$$\tau = \frac{\Delta P}{L}\frac{H}{2} \qquad (9\text{-}47)$$

(see also eq. (9-3) for τ_{wall}), where ΔP is the pressure gradient, L is the distance between the measurement points, and H is the slit height. The apparent shear viscosity was calculated with the power law (eq. (9-28)). The slit rheometer employed was a Haake model, and had a circular entrance. This model had to be attached to a figure-eight-shaped screw barrel for on-line apparent viscosity measurements. A transition piece was inserted between the two instruments to avoid vortexing and to provide a "smooth hydrodynamic transition" from the extruder into the die entrance (van Lengerich, 1990). If the die pressure becomes too high, the buildup of back pressure in a twin-screw extruder would cause plugged flow conditions (see Fig. 9-5); that is, the flow stops and the extruder fills quickly and completely with compressed food dough, causing a high-torque shutdown. The transition piece would prevent this undesirable back-pressure buildup by allowing smooth forward flow into the slit rheometer. The shear stress at the rheometer wall was plotted versus the angular velocity (or rotation rate) of the screw. The flow curves thus obtained for hydrated wheat starch (Fig. 9-12) were somewhat similar to those for hydrated maize starch (Figs. 9-9 and 9-10), and show clearly the shear-thinning behavior expected for pseudoplastic, power law fluids. As expected, shear-thinning effects increased at the lower water contents. To keep the twin-screw extruder filling at a constant rate, the screw was adjusted proportionally to the throughput rate. This methodology allowed for shear-rate variation in the extruder without changing the degree of fill in the twin-screw extruder.

The power law coefficients a and b for the data in Figure 9-12 are listed in Table 9-2. The values of the consistency coefficient, a, are very high and increase rapidly with the solids content, as expected. The values of b, on the other hand, are close to 1.0 for the starch samples containing 40% and 35% water, and are less than 0.5 below 30% water content. There is a sharp transition in behavior between 35% and 30% water content, which agrees with the observed hydration behavior reported by deuterium NMR (Yakubu, Baianu, and Orr, 1991). However, note that, as in most extrusion studies, the flow curves (Fig. 9-13) were analyzed only at relatively high shear rates (above ~ 200 s^{-1}), and therefore the b parameter values, and especially the a values, may not have been accurately determined.

The effects of extruder barrel temperature on the flow behavior of hydrated

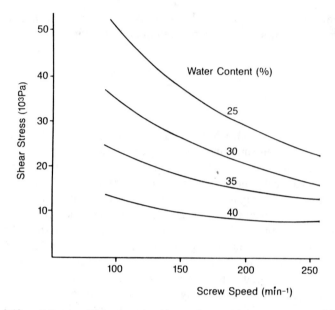

FIGURE 9-12.   Influence of screw speed on shear stress at various water contents (shear rate: 360 s^{-1}) (from van Lengerich, 1990, with permission).

TABLE 9-2   A Comparison of the Coefficients Obtained When Using Different Rheometers

Rheometer	ϵ/K	$\beta/10^{-3}$ $(\text{J} \cdot \text{g}^{-1})$	α	K_0 $(\text{Pa} \cdot \text{s}^n)$
Capillary	4960	—	−12.1	0.672
Extruder-fed	4820	−1.24	−26.8	80.2
Rheoplast	4250	−1.11*	−10.6	7.36

Rheometer	Power Law Index, b		
	$m_w = 0.21$	0.28	0.35
Capillary	—	0.40	—
Extruder-fed	0.31	0.48	0.68
Rheoplast[†]	0.37	0.41	0.44

*Density of melt assumed = 1260 $\text{kg} \cdot \text{m}^{-3}$
†At appropriate T_0.

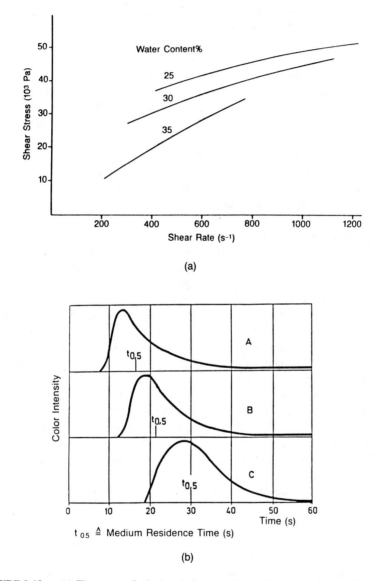

FIGURE 9-13.   (a) Flow curves for hydrated wheat starch at various water contents (from van Lengerich, 1990, with permission). (b) Influence of screw configuration on residence-time distribution (from van Lengerich, 1990, with permission).

wheat starch was also investigated, and the results were somewhat different from those of Parker et al. (1990) that were discussed in the previous section. The b value was highest for the highest barrel temperature at $m_w = 0.25$, and at 200°C the behavior was apparent-Newtonian, slightly expanded ($b = 1.05$) at screw speeds from 75 to 250 rot/min (apparent shear rates of ~360 s^{-1}).

These observations were explained as follows: the decrease in the apparent viscosity of the hydrated wheat starch at temperatures above 100°C causes the torque at the screw shafts to decrease, thus reducing the SME input (eq. (9-44)) and, therefore, decreasing the breakup, or degradation, of the extruded starch granules (van Lengerich, 1990). Thus, the flow behavior at 200°C becomes apparent-Newtonian (Fig. 9-14). Furthermore, the expansion ratio of extruded starches increased at a constant shear rate at lower water content. Extremely high shear rates (> 1000 s^{-1}) were thought to predominantly cause the molecular disruption of amylopectin molecules, which would keep the expansion ratio from increasing. The breakdown was more extensive at high screw speeds, but the residence time did not have a significant effect on the molecular breakdown of wheat starch in the extruder. The average molecular weights decreased with increasing SME values. Such average molecular weights were estimated first by dispersing the wheat starch in an aqueous dimethyl sulfoxide (DMSO) solution and then by using high-pressure gel permeation chromatography on porous glass to separate the dissolved large polysaccharide components of the starch. The elution profiles were calibrated with synthetic amyloses of known molecular weights. Two major peaks were observed by this technique, a dominant peak of M_w around 10^7 and a broader, shorter peak of M_w around 10^6. The first peak was attributed to amylopectins. The first peak decreased, with increasing SME, while the second peak increased (Fig. 9-15), presumably by breaking the glycosidic bonds in amylopectin molecules that would cause fragmentation of amylopectin. Large amylopectin fragments would correspond to intermediate peaks between the first (amylopectin) peak and the second, predominantly amylose peak (van Lengerich, 1990). The dependence of the average molecular weight, M_w, of the wheat starch extrudate on the SME input was reasonably well determined by the equation:

$$\overline{M}_w = \frac{MG_0}{1 + \exp\left(-\alpha \cdot \beta + \beta \cdot \text{SME}\right)} \qquad (9\text{-}48)$$

as shown in Figure 9-16. Above an SME value of 200 Wh/kg, however, the $\overline{M}_w$ became constant, suggesting that no further molecular degradation would occur above this SME input level (van Lengerich, 1990). The iodine binding test of the amylose from the wheat starch extrudates suggested that, unlike amylopectin, the amylose molecules are not degraded by extrusion to a significant

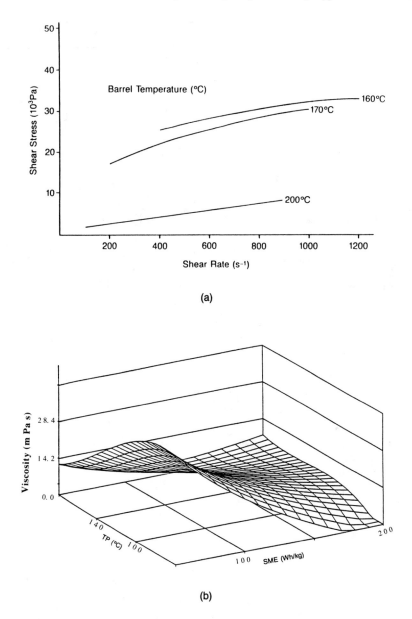

FIGURE 9-14.   (a) Flow curves of hydrated wheat starch pastes at various barrel temperatures (from van Lengerich, 1990, with permission). (b) Influence of the introduction of energy on the hot-paste apparent viscosity of extruded wheat starch (from van Lengerich, 1990, with permission).

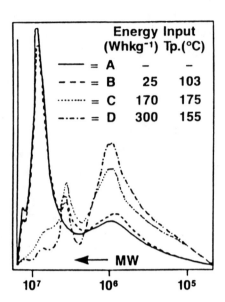

FIGURE 9-15. Influence of the introduction of energy on the relative molecular-weight distribution of wheat starch extruded under various extrusion conditions characterized by SME and T, P (from van Lengerich, 1990, with permission).

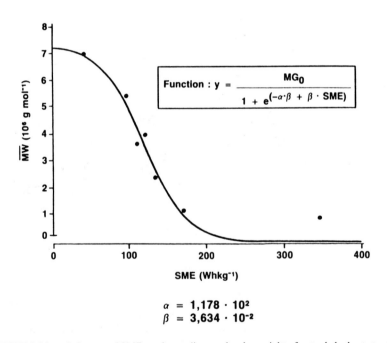

$$y = \frac{MG_0}{1 + e^{(-\alpha \cdot \beta \ + \ \beta \ \cdot \ SME)}}$$

$$\alpha = 1{,}178 \cdot 10^2$$
$$\beta = 3{,}634 \cdot 10^{-2}$$

FIGURE 9-16. Influence of SME on the medium molecular weight of extruded wheat starch (from van Lengerich, 1990, with permission).

extent (the iodine binding capacity remained at ~5.3% up to an SME of 343 Wh/kg and temperatures up to 175°C).

Upon heating, starch forms viscous pastes in water because its polysaccharide components are hydrated following the disruption of their hydrogen bonds and the formation of new hydrogen bonds with water in native starch.

Increasing the internal energy of the hydrated wheat starch during extrusion caused several changes in the starch structure, in addition to those related to gelatinization. Such structural changes were characterized by X-ray diffraction (Fig. 9-17), scanning electron microscopy (SEM), high-pressure gel chromatography (Fig. 9-16), the iodine binding test, and DSC (Fig. 9-18). The X-ray diffraction patterns of the hydrated wheat starch extrudates suggested a marked loss in the starch granule crystallinity at a very low SME input. Furthermore, hydrated wheat starches extruded at an SME of 80 Wh/kg and at 90°C had negligible, residual gelatinization enthalpy as determined by DSC. It was suggested that at higher SME inputs new crystalline complexes are formed between preexisting starch lipids and the amylose portion of the wheat starch. The SEM

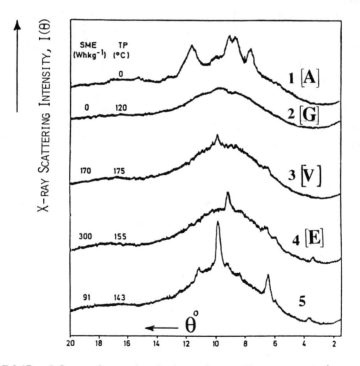

FIGURE 9-17.   Influence of energy introduction on the crystalline structure of wheat starch extruded under various extrusion conditions characterized by SME and T, P (from van Lengerich, 1990, with permission).

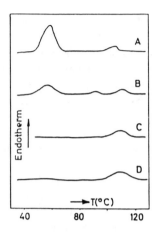

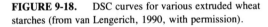

FIGURE 9-18.   DSC curves for various extruded wheat
starches (from van Lengerich, 1990, with permission).

pictures were interpreted as evidence that extrusion causes the wheat starch
granules to break up. At high SME input values the extrudate's apparent viscosities decreased in relation to the molecular degradation/partial depolymer-
ization of wheat starch amylopectins (van Lengerich, 1990). The SME was
linearly correlated with the shear stress in the slit die (Fig. 9-19) at a constant
degree of fill of the twin-screw extruder, up to SME values of 180 Wh/kg.
Therefore, the shear stress at the slit die wall can be used as a good indicator
of the SME input to the hydrated wheat starch system. It would be interesting
to see if this correlation also holds for wheat flour and corn dough extrusion,
as well as soy protein. Such measurements might provide a basis for developing
an improved understanding of the food extrusion process.

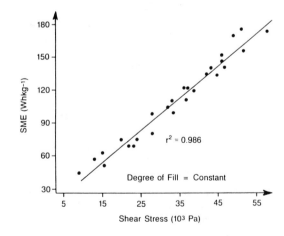

FIGURE 9-19.   Correlation
between SME and shear stress in
the slit die (from van Lengerich,
1990, with permission).

Although several techniques were employed to determine material changes caused by extrusion, the results that are available so far do *not* provide a complete characterization of the extrusion-induced changes. Among the techniques that are likely to make a significant impact on the extrusion problem are NMR and FT-IR. Both techniques are likely to provide a substantial amount of new structural and dynamic information that is directly relevant to understanding changes in the structure and hydration of the food materials during extrusion. Further computational and modeling efforts are also necessary to analyze non-isothermal, non-Newtonian flow conditions that have so far eluded rheologists. Such developments are now in press (Kokini, Ho, and Karwe, 1992); especially relevant are chapters 8, 10, 13, 20–29, 37, and 39 in this recent food-extrusion book that expand on the aspects outlined in this chapter.

References

Bagley, E. B. 1954. End-correction in capillary flow of polyethylene. *J. Appl. Phys.* **28**:624.

Baianu, I. C., et al. 1990. Multinuclear spin relaxation and high-resolution NMR studies of food proteins, agriculturally important materials and related systems. In *NMR of Biopolymers*, J. Finley et al., eds. New York: Plenum.

Boydson, J. A. 1981. *Flow Properties of Polymer Melts.* London: George Godwin.

Han, C. D. 1971. The effect of temperature on the elastic properties of polymer melts. *Polym. Eng. Sci.* **11**:205–210.

Han, C. D. 1981. *Multiphase Flow in Polymer Processing.* New York: Academic Press.

Harper, J. M. 1981. *Extrusion of Foods*, vols. 1 and 2. Boca Raton, Fla.: CRC Press.

Kakalis, L. T., and I. C. Baianu. 1988. Oxygen-17 and deuterium nuclear magnetic relaxation studies of lysozyme hydration in solution: Field dispersion, concentration, pH/pD, and protein activity dependences. *Archiv. Biochem. Biophys.* **267**:829–841.

Kokini, J. L., C.-T. Ho, and M. V. Karwe. 1992. *Food Extrusion Science and Technology.* New York, Basel, and Hong Kong: Marcel Dekker.

Lai-Fook, R. A., A. Senouci, A. C. Smith, and D. P. Isherwood. 1989. Pumping characteristics of self-wiping twin-screw extruders—A theoretical and experimental study on biopolymer extrusion. *Polym. Eng. Sci.* **29**:433–440.

Mercier, C., R. Charbonniere, J. Grebant, and J. F. de La Guérivière. 1980. Formation of an amylose-lipid-complex by twin-screw extrusion cooking of maniok starch. *Cereal Chem.* **57**:4.

Meuser, F., and B. van Lengerich. 1984. *System Analytical Methods for the Extrusion of Starches Thermal Processing and Quality of Food*, P. Zeuthen, ed., 175. London and New York: Elsevier.

Mora-Gutierrez, A., and I. C. Baianu. 1989. ^{1}H NMR and viscosity measurements on suspensions of carbohydrates and starch from corn. Relationship to oxygen-17 and carbon-13 data. *J. Agric. Food Chem.* **37**:1459–1467.

Mora-Gutierrez, A., and I. C. Baianu. 1991. Carbon-13 nuclear magnetic resonance studies of chemically modified waxy maize starch, corn syrups, and maltodextrins. Comparison with potato starch/maltodextrins. *J. Agric. Food Chem.* **39**:1057–1062.

Morgan, R. G., J. F. Steffe, and R. Y. Ofoli. 1989. A generalized viscosity model for extrusion of protein doughs. *J. Food Process.* **11**:55–78.

Parker, R., A.-L. Ollett, R. A. Lai-Fook, and A. C. Smith. 1990. The rheology of food "melts" and its application to extrusion processing. In *Rheology of Food, Pharmaceutical and Biological Materials with General Rheology*, R. E. Carter, ed., 57–73. London and New York: Elsevier.

Paton, Y. B., et al. 1974. Extrusion. In *Processing of Thermoplastic Materials*, E. C. Bernhardt, ed. Huntington, NY: Robert E. Krieger.

Rabinowitsch, B. 1929. Uber die Viskositaet and Elastizitaat von Solen. *Z. Phys. Chem.* **145**:1.

Renisen, C. H., and J. P. Clark. 1978. A viscosity model for a cooking dough. *J. Food Process. Eng.* **2**:39–64.

Rogers, M. G. 1970. Rheological interpretation of Brabender Plasticorder (extruder head) data. *Ind. Eng. Chem. Process Des. Der.* **9**(1):49.

Senouci, A., and A. C. Smith. An experimental study of food melt rheology. 1. Shear viscosity using a slit die viscometer and a capillary rheometer. *Rheol. Acta* **27**:546–554.

van Lengerich, B. 1990. Influence of extrusion processing on in-line rheological behavior, structure and function of wheat starch. In *Dough Rheology and Baked Product Texture*, H. Faridi and S. M. Faubion, eds., 421–471. New York: Van Nostrand Reinhold.

Vergnes, B., and J. P. Villemaire. 1987. Rheological behavior of low moisture molten maize starch. *Rheol. Acta* **26**:570–576.

Vergnes, B., and J. P. Villemaire, P. Colonna, and J. Tayeb. 1987. Interrelationships between thermomechanical treatment and macromolecular degradation of maize starch in novel rheometer with pre-shearing. *J. Cereal Sci.* **5**:189–202.

Yacu, W. 1985. Modelling a twin screw co-rotating extruder. *J. Food Process. Eng.* **8**:1–21.

Yakubu, P., I. C. Baianu, and P. H. Orr. 1990. Unique hydration behavior of potato starch as determined by deuterium nuclear magnetic resonance. *J. Food Sci.* **55**:458–461.

Yakubu, P., I. C. Baianu, and P. H. Orr. 1991. Deuterium nuclear magnetic resonance of potato starch hydration. In *Water Relationships in Foods*, H. Levine and L. Slade, eds., 585–597. New York: Plenum.

Yakubu, P., et al. 1991. ^{1}H and ^{2}H NMR studies of potato, corn and wheat starch gelatinization. Abstract No. 564 in *Proceedings of the IFT National Meeting*, A. M. Schenk, ed., 223. Dallas: Institute of Food Technology.

Index